Walberto Ortiz Sevillano

Nicknames, nicknames and nicknames in the Tumaqueña Speech

Walberto Ortiz Sevillano

Nicknames, nicknames and nicknames in the Tumaqueña Speech

Nicknames, identity and culture

ScienciaScripts

Imprint
Any brand names and product names mentioned in this book are subject to trademark, brand or patent protection and are trademarks or registered trademarks of their respective holders. The use of brand names, product names, common names, trade names, product descriptions etc. even without a particular marking in this work is in no way to be construed to mean that such names may be regarded as unrestricted in respect of trademark and brand protection legislation and could thus be used by anyone.

Cover image: www.ingimage.com

This book is a translation from the original published under ISBN 978-613-9-40909-9.

Publisher:
Sciencia Scripts
is a trademark of
Dodo Books Indian Ocean Ltd. and OmniScriptum S.R.L publishing group

120 High Road, East Finchley, London, N2 9ED, United Kingdom
Str. Armeneasca 28/1, office 1, Chisinau MD-2012, Republic of Moldova, Europe
Printed at: see last page
ISBN: 978-620-7-91179-0

JUSTO WALBERTO ORTIZ SEVILLANO

NICKNAMES, NICKNAMES AND NICKNAMES IN TUMAQUEÑA SPEECH

NICKNAMES, IDENTITY AND CULTURE

MULTIDIVERSE CULTURAL AND FOLKLORIC PHOTOGRAPHY

The Colombian Pacific is one of the largest and most biodiverse territories in the country. It is inhabited by Afro-descendants and indigenous people from different groups, but has little or no relevant and inclusive education. Here, the aim is to reflect on the importance of ethno-educational work that starts from the speech and oral tradition of the Afro-Pacific culture from the basic school, showing the obvious importance of endogenous ethnopedagogical work to achieve the cementing of ethnic and cultural identity in the early stages of academic life.

PRESENTATION

This paper approaches from a sociolinguistic perspective the study of nicknames in the speech of people from different social strata who live together in neighbourhoods belonging to Comunas (the political organisational structure of the municipality) and of different educational levels in the city of Tumaco. Our interest is focused on clarifying why people from different social strata currently use a "marginal or sub-standard language", which has traditionally been used to characterise and stigmatise others who live and interact in the same habitat in the city, regardless of whether or not they belong to their social stratum.
.This is based on the expressions and statements used by these people from different strata and in different socio-discursive spaces. Likewise, we analyse the nature of their nominalisations, the circumstances and the environments in which they are made, in which we find that such forms of nominalising or nicknaming, motear and sobrenombretear, arise in other types of situational scenarios characteristic of the marginal sectors of the city of Tumaco.This work demonstrates, with a data base obtained in the neighbourhoods of the five (5) selected Communes, that the use of para- languages is increasingly in vogue, fashionable, in use and in constant reference, not only by the lower social strata but also by upper social strata, whose expressions or nicknames in use, have been increasingly permeated by ways of nominalising, over-naming and nicknaming from a sub-standard, everyday and almost marginal, if not total, use.
Likewise, we show evidence of the linguistic and discursive differences that exist between people in the social strata to which they belong when it comes to giving nicknames. This, from the observation of their ways of life, customs and rituals of discourse and action. It should be noted that this research defines the different positions that are given in terms of the ways of naming, over-naming, nicknaming and, if you like, "ridiculing the other" based on the linguistic-marginal attitudes of the people of the city of Tumaco and their repetitive uses; all of this as part of the colloquial and anecdotal daily discourse that they tend to use in different areas of discursive interaction. In short, we analyse the uses of "sociolectal language" or sub-standard language that are generated in the speech of people of different ages, from different Communes and different social strata. Finally, we present the linguistic corpus obtained which, in an explicit and interpretative way, reveals the ways of naming another subject (group, academic, social pair) used by people today. We also present, through analysis and interpretation, the associations, lexical,

morphological and semantic recurrences that emerge therefrom and that people tend to use in their everyday life and in their processes of gestation of meaning and anecdotal sense.

New view, once touristy Morro Arch

INTRODUCTION

In Colombia, especially since the foundation of the Caro y Cuervo Institute (1942), and more specifically with the creation of its Department of Dialectology (1947), together with the interest and assiduous work, in many cases silent but effective, of national and international researchers, linguistic studies have achieved a great development and a very important place in the Hispanic and world academic concert.

Undoubtedly, up to this point, much has been done and achieved. We have come a long way, researching, collecting and analysing the living reality of our national language in its geographical and socio-cultural context[12]. This allows us to say with pride that Colombia is one of the Hispanic countries that has studied the language that today is spoken by more than 400 million people in the world the most.

The most characteristic aspect of the inhabitants of the Pacific is the dialect, in which many specificities stand out from the different points of view of the linguistic variation starting from the lexical-semantic to the morphosyntactic and phonetic-phonological ones among others that are reflected in the clearly marginal character, surrounded by jungle and sea that escape more and more to show the Spanish spoken in the Pacific and are expressed from there for Colombia and the world. The Afro-Pacific oral traditions are only a pretext to show the urban-rural movement of the Afro-Colombian culture that are seen and felt with each of its endogenous and exogenous changes expressed through the oral tradition and its passage to the written text in prose or verse as an important element of the black presence in the region that little by little is nationalised in Colombia and internationalised in the whole world.

With the Afro-Colombian oral tradition expressed in poems, verses, sayings, proverbs and compiled couplets, in riddles, they show that Afro-Colombian culture is transported from Afro to Afro and even to other ethnic groups, and also manages to bathe and cover them with the mantle and the stoic joy of everyday life that characterises the African diaspora within the national and international diversity. With the oral tradition one dances and communicates more. This reflection shows how the triadic relationship between Afro-Colombian oral tradition (fundamental element of speech), memory and ancestral knowledge appear and disappear in the spontaneous communication of the pacificans to show how, despite the difficult situations known to all, Afro-Colombian expressions of identity still persist and how they go over the wall of lamentations to become bearers of many joys that relax the soul and make the spirits rest. Also, mother of sayings and proverbs, bearer of myths and legends, maker of verses and couplets, builder of décimas both divine and human, depository of the collective memory of the Pacific region is the moulder of the childish and youthful behaviours of many of us. She allows us to teach and learn, to recreate and play with rondas and riddles. In addition to understanding the world from the Afro-Colombian point of view.

JUSTIFICATION

The present work approaches from a sociolinguistic perspective the study of the nickname, nickname or nickname in the speech of different social strata settled in the five communes, of different socio-economic strata of Tumaco, a port on the Colombian Pacific coast. Our interest is focused on clarifying why young people from higher social strata currently use a "marginal or sub-standard language", which has traditionally characterised and stigmatised another sector of the city's young people from lower social strata.
This is based on the different neologisms, expressions and statements used by young people from the upper stratum in different socio-discursive spaces. Likewise, we analyse the nature of their nominalisations and the environments in which they are used. We find that such forms of nominalising or nicknaming arise in other types of situational circumstances that are characteristic of the peripheral, vulnerable sectors of Tumaco.
In this work we show, on the basis of data obtained in the chosen community, that the use of anti-languages is increasingly in vogue, fashionable, in use and in constant reference, not only by the lower social strata but also by the higher social strata, whose language in use has been increasingly permeated by ways of nominalising, over-naming and nicknaming from a sub-standard, everyday and almost marginal, if not total, use. In the same way, we show evidence of the linguistic and discursive differences that exist between adults, young people and even the elderly in the different social strata to which they belong when it comes to using names. This, from the observation of their ways of life, customs and rituals of discourse and action. It should be noted that this research defines the different postures that are given in terms of the ways of naming, over-naming, nicknaming and, if you like, "ridiculing the other" based on the linguistic-marginal attitudes of the young people of the communities of Tumaco and their repetitive uses; all of this as part of the daily community discourse, which they tend to use in different areas of discursive interaction. In short, we analyse the uses of the language of the periphery and of sub-normal or sub-standard neighbourhoods that are generated in the speech of people of different ages, average of the same gender (male-female), but from different social strata, inhabitants of. Finally, we present the linguistic corpus obtained which, in an explicit and interpretative way, reveals the ways of naming another subject (group, academic, social pair) used by people today. We also present the associations, lexical, morphological and semantic recurrences that emerge from the analysis and interpretation, and which people tend to

use in their everyday life and in their processes of gestation of meaning and linguistic geobarrial sense.

Preliminary notes.

This research study is part of sociolinguistics as an interdisciplinary discipline oriented towards the socio-linguistic phenomena of speech and its presence and/or recurrence in the linguistic communities in which it occurs. In its development, it shows, from the analysis of speech in the Tumaqueña community, the way in which certain forms of pejorative enunciation are used to highlight the physical, mental, spiritual, psychic or moral characteristics of people, groups of friends, academic peers, etc., especially those of the same age range, who share discursive spaces and similar social representations of the world of life that constructs them.
In this sense, with the aim of unveiling this phenomenon of the speech of the communities settled in the neighbourhoods that belong to the 5 communes of Tumaco, we start from a series of samples that have been collected in the chosen linguistic community, which shows the daily discursive uses in the diverse conversations in which they engage. These samples have been taken from the different socio-economic strata of the city and are the result of their own participation in a process of identity and rootedness which implies closeness to the subjects under study. This means that the present research, being a n approach to something so close to our linguistic-discursive reality, enables a close and faithful perspective on the sociolinguistic phenomenon of the nickname assumed by young people in the city today. The "corpus" of this work, as has been pointed out, has been collected over two years of the year two thousand and eight (2012-2014) of which - it must be said initially - contains elements that can easily be seen as a ridiculing speech act: without it specifically being the speech act ridiculing, as it should be remembered that the samples mentioned here are part of various speech acts and not of one in particular. The user of the nickname, "mote", "sobrenombre" normally uses it with colleagues, friends and family, but it is also recurrently used with strangers, although the latter is increasingly less common, largely due to the distances taken by the different socio-economic strata, despite being inscribed in the social and democratic dynamics of the 21st century, where apparent equality is nothing more than a hopeful challenge.In the same way, a very strong influence of the so-called "marginal languages" can be glimpsed, which are not only used in their great majority by the subjects belonging to the lower social strata of the city: it seems that these languages are increasingly impregnating

the young people of the middle social strata, and in turn, the speakers of the upper social strata, who are resorting to this type of speech, of recursivity in intercommunicative events. The cause is not very clear; it could be indicated that this happens because of imitation of the environment, because of the closeness between young people belonging to polar social strata, or just because it is assumed as a social fashion or tendency towards the use of pejorative terms that can become offensive according to the situation and the environment in which they are manifested. It is this force that motivates the present case study in order to show how and why the upper social strata use marginal discourses on certain occasions either by stimuli that trigger effusiveness, ridicule or anger in their everyday linguistic realisations.

Hence, it is very complex to establish the origin of the use of the nickname in the Spanish language, due in the first instance to the heterogeneous nature of Spanish, which incorporates a multitude of native American languages in its journey of more than three centuries, since the vernacular languages as well as the culture they transmitted were preserved in some linguistic components such as lexical, adapting to the Spanish idiosyncrasy and sporadically sprouting in the most sincere (almost unconscious) actions that reveal a phylogenetic potential of the nickname.

In any case, it could be said that it is typical in Latin America that the nickname serves as a bridge that unites the different othernesses in social interaction. This is the legacy of a class-based division of society; nowadays it is economic power in families that seems to determine the course of social relations which, without falling into a Manichean judgement of the use of the nickname, can transcend the way in which we become aware of our position in the speech community, in society and in the world.

The nicknames, nicknames or nicknames, broadly speaking, are ways of enunciating the other subject. It is Tumaco, a municipality in the Nariño department of Colombia, where we can call, from the dynamics of the speech of the people of the neighbourhoods who live in the communes, "the classification of what is named "2 , because the perceptions of the discursive environments show that there are already in our Tumaco context, three types or ways of naming someone: the first has to do with what is "the nickname" (as that name given to a subject because of its resemblance or visual equality with what is referred to or stated as a nickname), the second is "overnaming" (assumed as that way of naming in which there is a pejorative sense), which depends on the issuer, However, in all cases of nicknames there is no desire to attack the other, but rather, what is sought is to question in an abrupt way, perhaps releasing tensions in order to produce an immediate effect of closeness with the interlocutor.It is also the way in which we refer to the violent characters in our society, in our case in the Tumaqueño environment: "the alias" which is assumed as that re-naming that the subject, the individual or the young person deserves for their violent, criminal characteristics and, in the best of cases, for the associations that in youth discourses are generated without corresponding to the delinquent or punishable reality.

It is only logical to think that our city has become a discursive space for young people that shows linguistic marks that are usually manifested at the moment of enunciation in the various city and youth spheres. Within the different speech acts that make up a communicative event in spontaneous conversation, the everyday life of young people is always exposed to the use of the variables of language that unfold in their social nature. This is what undoubtedly motivates us to carry out the present study from a synchronic perspective and thus explain the phenomena. sociolinguistic aspects of the transformation of the linguistic performance that the speech of the Tumaqueño community is undergoing today.
Therefore, it is understood that the population, the object of the research, in their communicative and discursive rituals tends to emphasise the differences and psychological or physical characteristics of each subject (academic, group or social), being then this community, discursive subjects that tend to classify, specify, determine or ultimately ridicule the other members of their youth, group or sub-cultural community. This, to the extent that the young person tends to frame others depending on their physical, psychological characteristics or their worldviews.

In short, to give an account, on the basis of a study, of the universality of nicknames, nicknames or nicknames and their survival in Afro-descendant neighbourhood societies organised in Communes, as well as their daily use as identifying elements and as a convivial link. Describe from a communicative, expressive and creative point of view the rich intangible heritage of Afro-descendant societies manifested in nicknames. To construct a corpus based on the records made in the communal neighbourhoods of Tumaco.

THEORETICAL FRAMEWORK CONCEPTUAL

In view of the fact that this work refers to a language component such as the lexicon of the Venezuelan adolescent, it is necessary to divide the theoretical framework into two fundamental aspects. Firstly, some definitions of lexicon and the role it has been assigned in linguistic studies will be analysed. Some characteristics of the adolescent lexicon will also be briefly discussed. Secondly, reference will be made to word formation procedures and relevant theoretical aspects related to this topic that have been dealt with by different linguists. Among the mechanisms used in word formation, emphasis will be placed on nominal composition. The importance of context in the production of nicknames will also b e mentioned.

The study of the lexicon has been of interest to different specialists, who have analysed it from different perspectives. This fact has given rise to discussions, as can be seen in the extensive bibliography on the subject. For example, the semanticists approach the problem from the point of view of meaning and the structuring of the lexicon into semantic fields. On the other hand, other linguists study it from the point of view of grammatical phenomena. However, both agree on the social and cultural importance of this component, as Sapir (1974) states: "the vocabulary of a language is the one that most clearly reflects the physical and social environment of its speakers" (p.21). These ideas will be taken up further below.

In the following, it is considered necessary first of all to present several concepts of lexicon that will allow us to situate ourselves in the problem that will be developed in this work. According to the Diccionario de la R.A.E. (1984), the lexicon is: "vocabulary, the set of words of a language, or those belonging to a region, a given activity or a given semantic field" (p.28). On the other hand, Seco (1999) considers that the lexicon is a set of words belonging to a given region, activity, human group, work or person" (p.2823). For Mounin (1979), the lexicon is a set of significant units of a language at a given moment in its history. There are major coincidences between these authors, only Seco adds the possibility of speaking of the lexicon of a work such as Canaima, Doña Bárbara and of a specific person (lexicon of a famous person, for example, Arturo Uslar Pietri). Mounin, for his part, stresses the importance of the historical changes that are inherent in the lexicon of a region. We agree with all of the above. It is clear that, within Spanish, one can speak of the lexicon of the Spanish spoken in Venezuela, or of the lexicon of the Andean region, as well as of a specific activity, such as the speech of doctors, engineers and architects.The changes that occur in the lexicon from a generational point of view should also be highlighted, an example of which is the difference between the lexicon of adults and that of adolescents. The lexicon of the latter age group is the subject of this study.

After reviewing these concepts of lexicon, it is necessary to review some linguistic approaches that have given an account of this component of language. Martinet (1969) says that although structuralist linguists have insisted on the systematic character of languages, many of them deny this character to the lexicon altogether. In the case of North Americans, at least, lexicography lived apart from structuralist theories. Nowadays, on the contrary, the aim is to construct a structural lexicology, i.e. a theoretical study of the lexicon, attributing to it a structured character. For Martinet (1969), the lexicon, like grammar, deals with the units of first articulation or monemes. One could call lexical all the monemes that appear as independent articles in ordinary dictionaries, i.e. what we commonly call words, including units such as prepositions and conjunctions. In linguistics, however, in order not to dissociate elements of analogous function, such as prepositions and desinences, lexical units and grammatical units are opposed to each other. Lexical units are generally considered to belong to unlimited or open inventories, while grammatical units belong to closed limited inventories.

While traditional lexicology studied words in theoretical isolation, recent attempts in lexicology and structural semantics are not limited to the study of lexical fields, but also try to determine the latitudes of semantic combination of lexical units in the sentence.

For Martinet (1969) lexicology is part of linguistics, if it aims to account for the verbal reactions of interlocutors communicating with each other. Although its methods have not always been linguistic, this is due to the fact that the lexicon is the level of language which emerges most easily into the consciousness of speakers, because it is directly related to meaning and is closely linked to cultural evolution. Hockett (1976), who belongs to the structuralist current, considers that describing a language means analysing what he calls the grammatical nucleus, since this constitutes the formal schema of the language, which is acquired by speakers from a very early age and once acquired remains unchanged over time. Thus, only the lexical component will change, which is responsible for filling in the final part of this schema with words. That is to say, to place the nouns that correspond to the nominal syntagm (SN), the verbal syntagm (SV), the noun (S), the verb (V), the adjective (Adj). While there are few changes in the grammatical nucleus, this is not the case with the lexicon, which adapts to the expressive needs of speakers, to technological advances, to social changes and generational changes.Mounin (1979), referring to Generative Grammar, points out that the lexicon is a subcomponent which, together with the categorial subcomponent, forms the basis of the syntactic component. It consists of an unordered list of lexical units and also comprises a number of redundancy rules. Lexical units are associated with substitution transformations that insert such units into positions marked by the occurrence of complex symbols within the strings generated by the categorical component of the grammar.

Abraham (1981) on the same aspect, considers that the lexicon records, in principle, all the lexical units of a language and associates with them the syntactic, semantic and phonological information required for the correct functioning of the rules of sentence structures (p.125).

These ideas show that for both generativist and structuralist authors, the lexicon is part of the grammar of a language. However, it should be noted that the authors cited agree with Martinet that the study of the lexicon belongs to linguistics, as has been pointed out.

From another point of view, Lamíquiz (1975), when studying the semantic component of language, considers it feasible to classify it in three orders: logical semantics, which develops a series of logical problems of signification, studies the relationship between the linguistic sign and reality, the necessary conditions for a sign to be applied to an object, and the rules that ensure an exact signification. Psychological semantics, which tries to explain why we speak, what happens in the spirit of the speaker, and in the spirit of the listener, when we communicate, what is the psychic mechanism that is established between the speaker and the listener. And thirdly, we have linguistic semantics, which deals with meaning within the communication system and describes how it works. Here it is necessary to pause a little to specify conceptually and terminologically its components. It is understood as belonging to this field, meaning, everything linguistic that refers to the study of semantics and lexicology. In other words, these terms: semantics and lexicology are closely related. In semantics, the lexicological unit or lexeme and the unit of meaning or semantem, have differences, but through an example we will refer to the linguistic relations between both units of different focus, which interrelated form the semantem or unit of meaning. Let's look at the following example: the form **guarapa**, which we find in bold, in the Diccionario del Habla Actual de Venezuela (1994), is located alphabetically in the letter **G**, it is the lexeme, a form studied by lexicology. Its complete definition, which we find in 2) A refreshing drink prepared with fruit juice, especially lemon or pineapple juice, lots of water and papelón. This definition is the sememe that is studied in the semantic functioning. The complete record of the term, that is, its lexicological form or lexeme and the semantic function or sememe, as a whole, interrelated, constitutes the semantem. In other words, for Lamíquiz, the lexicon is closely linked to semantics. In view of the historical controversy about whether the study of the lexicon corresponds to linguistics or to lexicography, this paper assumes that the lexicon can be studied from different perspectives, but without leaving aside the morphosyntactic levels of the language, as well as the pragmatic strategies used by speakers.

Nicknaming: a motivated speech act

Do you know what a nickname is? Have you suffered the anguish of having a nickname? Do you know what your friends call you? Do you like it? Does it offend you? Does it make you unhappy? There are many questions we could ask about this sociolectal linguistic phenomenon.

According to LOZANO RAMÍREZ, (1999), the nickname "is an act of creation or motivated linguistic recreation, very expressive, by means of which the subject giving a nickname gives a new name to his fellows, according to the characteristics that evoke in his mind the image of an object, thing, subject or circumstance and that identify the character that receives this new name". The speaker's constant need to designate realities leads the language user to awaken in his memory (brain) images that are ready to be activated, allowing the creation of this linguistic sign. The speaker has an immense associative capacity that works like a machine in pursuit of the nomination, e.g.: Correlón, Piangua, Cautiza, Caballo, Donsega.

Surely, no one escapes having a nickname. They do not come out of nowhere; they always have a motivation and this is the reason that leads the nickname-giver to unload on others the emotional force aroused by the characteristics of the subjects for various reasons: affection, love, hatred, anger, malice, jocularity, emotion, sadness, envy, friendship, enmity, etc. From ancient times to the present day, men have had nicknames. This is not a new phenomenon. What is new for linguistic creation are the characteristics, circumstances, socio-cultural conditions and intentions of people in the different epochs. Thus, the nickname as a motivated linguistic phenomenon is always tinged with the emotive, festive, humorous, sarcastic or rude colouring that the speaker who creates or recreates the sign gives it. Nicknames or nicknames are words that constitute a highly economic unit of discourse from a linguistic perspective. They synthesise a large amount of information, communicative intentions and convivial attitudes which are understood, above all, by the frequent users of them, as is the case of people in rural areas who maintain very close relationships of coexistence. They are the ones most capable of accurately decoding the meaning, senses and intentions of these appellatives according to the communicative situation, the context and other pragmatic variants, in addition to the purely semantic and prosodic ones. These appellatives are, together with others of a similar nature, synthetic discourses twinned under the hyperonym of nicknames.

Importance of context

In previous paragraphs, the importance of context in the creation of the lexicon was discussed and it was noted how, through this component of language, it was possible to appreciate not only linguistic phenomena, but also socio-cultural and psychological phenomena present in individuals and social groups. The importance of the communicative act in word formation was also stressed. It is necessary to emphasise that every verbal or written utterance originates from three types of contexts: psychological, social and situational. The first refers to the relationship that may exist between the participants in the communicative event. It has to do with the fact that verbal utterances are appropriate to the actions implicit in them: a request, an order, for example, entails the need for a certain relationship between speaker and listener. Psychological context is closely related to the shared knowledge between speaker and listener. This context relates to the set of assumptions that the sender intuits about the receiver. Social context is very complex, it has to do with social behaviour in various cultural circumstances, as has been pointed out: customs, traditions, history. Today, the social context is strongly influenced by the role of the media: newspapers, radio, television, advertising, internet. Today, they are perhaps the most powerful forces permeating language use. This is evident in the corpus collected in this paper, as many of the nicknames are taken from the media. The situational context includes in this case all those aspects outside the environment surrounding the communicative act that can in any way influence the mechanisms for its production and comprehension. That is to say, elements such as: the physical, ceremonial, formal environment, mood, joy, familiarity, contribute to making it possible for a communicative act to fulfil the task proposed by the sender.The concept of context is essential for all linguistic studies that will be approached from a pragmatic or discourse-textual perspective. Precisely, the aspect which most clearly defines this type of studies and, at the same time, distinguishes them from those carried out from a strictly grammatical point of view, consists in the fact that the former incorporate contextual data in the linguistic description. Halliday (1983) talks about the textual environment or context, i.e. the utterances surrounding what is being considered for analysis, since the concrete meaning acquired by words, utterances and discourses depends to a large extent on what has been said before and what comes after.

The creation and recognition of contextual factors are the fundamental aspects of human communication. Context is dynamic and those involved in a communicative exchange have to construct, create, change and interpret it. In this process, certain elements such as the physical and cultural environment and certain collective behavioural norms or tendencies internalised cognitively in the form of frames or scripts may occur. However, it is people, through the activities they carry out, who actualise these factors by making them a significant part of what is happening. Thus, in this study, it is adolescents with their lexicon who are constantly renewing their way of speaking, through the actualisation of different social and psychological contexts.

So what are nicknames?

The definition of the dictionary of the R.A.E. (1984) states, on the one hand, that: it is a name usually given to a person, taken from their bodily defects or some other circumstance, and on the other hand, it is a joke or funny saying with which a person is described. person or thing, usually by means of an ingenious comparison. This definition is adapted to the results found in the corpus of this study, as will be shown later on. In the same vein, Ledezma and Obregón (1989) state that the creation of nicknames by adolescents is a significant mechanism in the production of humour. They argue that the linguistic resources used in the creation of nicknames are varied, in some cases the process is simple, since it is a matter of identifying a physical characteristic with some aspect of reality. This can refer to animals or also to plants, for example **gorilón** (similar to a gorilla), **taparón** (similar to a tapara), in both cases the suffix **-ón** , gives it an emphatic and humorous nuance. Others, on the contrary, are complex in their structure because, in some cases, the nominal composition plays a very important role, for example: **zancadilla,** which is the fusion of zancudo with ladilla and which refers to a very annoying person. In the aforementioned research Ledezma and Obregón (1989) analysed only a small sample of the corpus collected in interviews with students aged between 14 and 16. The researchers selected those nicknames in whose formation complex processes are manifested that show the adolescent's creativity. This work constitutes an antecedent to the research being carried out, as it delves deeper into these complex processes of nickname formation by Venezuelan adolescents.The speaker uses the language's own resources in the nominative process for its creation. Thus, the speaker has two ways to elaborate his speech act. One is morphological and the other semantic. From morphology, nicknames are created: 1. by derivation (diminutives: Benitín, Cebollita; augmentatives: Carietón, Lagrimón; derogatory: Viejorro, Bojote, Pecueco; superlatives: Blanquísimo, Tontorrísimo). 2. by composition (Camaralenta, Cristoviejo). 3. by transplantation (simpson, Barbie, Brownie, Batman). From the semantic point of view, they are created by metaphor, metonymy, synecdoche; the subject is associated with animals, plants, fruits, objects, products, activities, work tools, professions or trades, religious life, human body, defects, virtues, etc.*.Nicknames are found in all social classes. In the upper classes, they are abundant, as well as in the middle and lower classes. According to these levels of language, there are nicknames ranging from the most original, noble, gentle, jocular or humorous, even the most grotesque, vulgar, satirical and denigrating, such as: Careniño, Niñodios, Bastantica, Patapicha, Virgojecho, Sabañón, Mocotieso, among others. We cannot and should not confuse the terms nickname (carecrimen, gusano, bacteria), name (Blanca, Juan, Andrés), nickname (el valiente, el manco), pseudonym (Dartagnan, Cleofás), hypocoristic (Chepe, Pepe, Pocho), alias (Sangrenegra, el Patrón) because, although they refer to the new name given to the subject, they differ in significant nuances. However, the terms apodo, mote, remoque or remoquete are equivalent.Thus, the nickname as a motivated linguistic phenomenon is always tinged with the emotive, festive, humorous, sarcastic or rude colouring that the speaker who creates or recreates the sign gives it.

Nicknames or nicknames

As we mentioned in the introduction, nicknames exist everywhere, in the same way that proper names or other types of nicknames appear with an identifying and denominative function. But it is also true that they are used much more in societies or human groups in which there is a close coexistence such as a village, a neighbourhood, a school, a workplace or a place of work that favours some kind of human contact between the people who share the work environment. In this way, they appear throughout the Spanish, European and Latin American rural world, as well as in the American and Australian indigenous cultures, to give some examples. We can say, therefore, that the use of nicknames is a universal phenomenon that responds to common social needs and conditions that favour their use. Ultimately, it is another socialising need that makes language adapt to the communicative needs of human beings, who look for utility and aesthetics in it.

The reason why nicknames appear is undoubtedly due to the fact that they have to cover a series of communicative, social, emotional and aesthetic needs. They serve to distinguish, identify, specify identification; to mark a relationship between some characteristics of the nickname and its nickname, as well as to establish ranks or social groups, to provide affective values, to introduce connotations, to classify,

To economise on language or gain relevance, to offend, to encourage and practice creativity, to establish comparisons, to play with language, to literaturise - there are certainly many reasons why nicknames arise and are used, and many functions they serve, although it is sometimes difficult to establish precisely which of them are the priority ones and which are secondary.[3]

Although at first sight it might seem that the decline of the rural world would mean the disappearance of nicknames, the reality shows that this is not the case, since nicknames survive and continue to be used. We can see this every day in everyday life. They are used in family, friendly, neighbourhood, work, sporting and educational relationships, whether in schools, institutes or universities, where not even the teaching staff are "spared". In short, nicknames survive and survive in very good health. Even if we take a look at literature, we can verify their validity; and if we do so in the press and the media, in sports or political circles, we can easily see that almost all people-public figures have their nicknames or nicknames, more or less respectful, accepted or rejected - and the attitude of manifest rejection of the nickname must be taken care of, because, if anger is felt when it is used, this nickname will be consolidated and this universality of the nickname confirms that it responds to a communicative need. That is why it survives quite naturally, as do all the communicative formulas that we use for coexistence. And so we can see that bullfighters, boxers, cyclists, artists, politicians and people in public life are all deserving of nicknames. Sometimes they are new nicknames that respond to the public's perception of them; sometimes they are the recovery of childhood nicknames that become public like the character himself; sometimes they are nicknames that respond to diverse and unforeseeable reasons and circumstances.

This is the reality of nicknames, of which the dictionary says that it is a "name usually given to a person, taken from his bodily defects or some circumstance, de suso. A joke or humorous saying used to describe a person or a circumstance, de suso. thing, usually by means of an ingenious comparison". (DRAE, 1992: 1 12)2. And we can find a whole semantic field around this type of appellatives: nickname, nickname, nickname, nickname, alias, remoquete, cognomento, agnomento, renombre, sobrehúsa, pseudónimo, alcuña, alcuño, malnombre. In short, a whole set of denominative terms in which nicknames or nicknames are inscribed and which are regularly based on the ingenious comparison of defects, characteristics or circumstances.

As can be seen, there are many nicknames that can be typified. However, if we look at nicknames in particular, it is useful to establish a common basis for identifying them. In our study on nicknames or nicknames, especially if they are consolidated or familiar, we established some requirements for them to be considered as full nicknames. Among them we can mention the following: I. They fulfil the appellative and distinctive functions, par excellence. 2. They last for a very long time and accompany the person they are nicknamed after practically all his or her life. **3. They are** transmitted hereditarily to the family or to some of its members. 4. They undergo a continuous process of de-identification4 . The identifying capacity of these nicknames is so great that some famous people survive in the collective memory because of their nickname-pseudonym, as can be seen in the following examples...

As we have already mentioned, we believe that Moliner's definition is the one that best fits our vision of nicknames or nicknames. The lexicographer defines them as follows:

nickname. "Mote. A nickname sometimes applied to a person, among ordinary people, and very frequently in villages, where it is handed down from father to son; mote. "Nickname. A nickname, usually alluding to some quality, likeness of the person to whom it is applied, by which that person is known. Especially those used in villages, which are passed on from father to son and are generally not taken as offensive (Moliner, 1998)[5]. This author also provides us with three important guidelines: 1, that they abound or are frequent in villages; 2, that they are passed down from parents to children; and 3, that they occur "among ordinary people "6 , although, "generally, they are not taken as offensive". To some extent, these are some of the principles that we understand to be the requisites for a term to reach the full category of nickname or nickname.

They fulfil the appellative, distinctive and social functions. 2. They last for a very long time and accompany the person with whom they are nicknamed practically for life. 3. They are passed on hereditarily to the family or to some of its members. 4. They undergo a process of continuous de-anticisation (Ramírez, 2003). However, it is important to clarify the meaning of these appellatives from the perspective of rural people, as they have expressed it to us in the different investigations we have carried out. The perception they have of them is as follows: 1. **Sobrenombre: a** term that is hardly known and without a clear linguistic location for most of them; after explaining it, it is felt as a fine and cultured term of nickname or nickname, not very profitable in the sociolinguistics of the people. 2. **Apodo: a** mild nickname, almost a euphemism for mote. 3. **Mote:** it is the pure term of the rural nickname, the most frequent and widespread as an appellative in rural environments. It is given to a person for different reasons, sometimes without pejorative intention, as a linguistic-expressive synthesis of a sign of identity, of an anecdote, of a complicity; but in others with a light, medium or strongly offensive intention: it is a clear identifier and, in many occasions, it is extended to his family6.

Nicknames and their synonyms, from our point of view, and after these incursions into the lexicographical field and into the field of meanings perceived in the researched environment, are enduring terms, words, syntagms, phrases or noun phrases or sentences which, often with a pejorative nuance, despite the opinion of some users, compilers and scholars who state that there is no intention to offend and that the nominees are not bothered, always identify people and, with often characterised by linguistic caricaturisation and a wide range of social and convivial motifs.

We believe that the perception that the true value of the nickname lies in its figurative and intentional meaning and not in the straight meaning of the term (Moreu Rey, 1981) is very accurate, except in the case of those that respond directly to a trade, to a proper name or to a surname; and we could say that, even in these cases, it ends up accumulating figurative and added values and meanings which, through the tone, indicate and connote some characteristics of the nicknames and the convivial relationship between the interlocutors. To summarise, it could be said that nicknames fall within the science of onomastics (anthroponymy), i.e. that which deals with the proper names of people. It

should be borne in mind that nicknames were the first proper names to undergo a gradual process of de-typification. The study of these names provides a great deal of information about beliefs, forms of social organisation and relations of treatment, treatment and coexistence.
These individualised personal names or nicknames, whether very formal (forenames and surnames) or informal (nicknames or nicknames), have been assigned and used differently from one culture to another, but the process has always been the same. First names, surnames and nicknames are the most frequent ways of identifying people, sometimes with a single term, and sometimes with a combination of them in order to distinguish between those named. It is worth remembering that the first proper name was the nickname, nickname or nickname, from which the official, administrative and legal proper name was derived; but, as has already been said, the first was the nickname. However, due to the official identification of people in today's societies, nowadays it is often the other way around, first the proper name is assigned and then the nickname is given; a good example of this is the process of creating nicknames in some Aboriginal communities in Australia (Morgan, 1991).
In any case, we must point out that, although the study of these appellatives is currently being studied in depth, much remains to be done, as many of the works on them are reduced to the mere collection and compilation of the nicknames that occur in some villages. It is a complex and difficult field to work in, among other reasons, due to their strong identifying character and the residual semantic value of many of them with a negative meaning, which can lead to problems, even legal ones, in the study and, above all, the publication of terms such as El Mierda, Basura, Matabuelas, Caraculo, etc. Nevertheless, there is an abundant bibliography (IDATP Commission, 1954 et seq.), studies from the 19th century (Godoy, 1871), some from the 20th century (Moneu Rey, 1981; Barrio, 1995), others from the 21st century (Ramos, Da Silva, 2002); (Mangado, Ponce de León, 2007) and some doctoral theses on these terms in recent times (González, 1993, Ramírez, 2003)[7].

Classification of nicknames

It is a fact that our societies are becoming colder and colder, that coexistence is becoming less and less human and endearing, that we are becoming almost automatons who move through the streets without even looking at those we meet. But, fortunately, this situation, which is very common in our densely populated cities, is not yet the case in the villages. Living in a village means relating, contacting, communicating, dealing with each other on a daily basis and, why not, getting to know each other by nicknames.Because the nickname is born deep down from a need to communicate, to contact with the other, to refer to the other, and therefore brings us closer and tightens the ties that in a small town, like a spider's web, bind us to each other.

Obviously, this phenomenon of nicknames is not exclusive to a particular village, but is part of the ethnographic heritage of all, and it would be very revealing and endearing to compile and classify them extensively at the level of a county or even the region. This would contribute greatly to defining an idiosyncrasy, a way of being. First of all, nicknames should be distinguished from surnames and nicknames. The difference with the surname is clear, as the latter is given to us generation after generation and has an official and administrative character and a universal and timeless validity.

Nicknames and nicknames, however, are only valid in our village and have, of course, no validity for administrative or bureaucratic purposes. Although the difference between them and surnames is clear, the distinction between the two concepts is not so clear. In my opinion, the nickname is born with a purely differentiating intention, it is passed on for several generations, becoming a clan reference. The nickname, however, is born with a pejorative purpose, originated by some negative condition, physical defect, and originally refers to a specific person. What happens is that, over time, it can become a nickname, covering a more or less extensive lineage.

The truth is that both are of tremendous variety and richness, reflecting ingenuity and, sometimes, what can be colloquially called bad ideas, especially the nicknames.

As this is the sphere in which I live my life, I am going to focus this subject on a specific town, my own, Abarán, the town of the waterwheels and the balcony of the Ermita, of the Park by the Segura and the Cervantes Theatre with walls dotted with zarzuela notes, of the kiss to the Child every 6th January and the parade of Gigantes and Cabezudos at the end of each September. Here, as in all villages, the value of the surname is hardly differentiating, as Gómez, Ruices, Carrascos, Torneros, Cobarros, flood the Civil Registry. And Joaquín Gómez Gómez Gómez, to give an example, there could be dozens. The same happens in Blanca with the surnames Cano, Molina, Núñez. . . or in Ricote with the surnames Torrado, Miñango, Candil...From this problem arose the need to add something to distinguish one Gómez from another, for example. Already in the 16th century, we find some differentiating appellative, which refers to the place where one lives, and so we speak of Ginés Gómez de la Plaza and Ginés Gómez de la Calle. Evidently, we are dealing with a very small village made up only of what is today its old quarter.

Moving on to the present day, based on an alphabetical compilation made by my friend Indalecio Maquilón de Jerines, making a tasting or selection among the hundreds of nicknames/ nicknames collected, we could try to group them into semantic fields in order to establish a certain order in such an abundant jungle. Although an even more curious and interesting task would be to investigate the reason for each nickname, something that in most cases is lost to us in the mists of time, as many of them are born of a specific occurrence or a very specific circumstance that is impossible to follow over time, as there is no oral or, much less, written record of them.

We will be content, therefore, with taking a pleasant stroll through this forest of appellatives, trying to carry out a certain classification from a semantic or, sometimes, phonetic point of view.

a) Animal names: eagles, rabbits, chickens, chicks, lizards, crickets, mice...

b) Names of trades (some of them no longer in existence): "alañaores", "talabarteros", "garbanceros", "quincalleros", "santeros", "morcilleros"...

c) Physical/moral defects or virtues: one-eyed, crippled, one-armed, one-eyed, one-eyed, sad, snitches...; good-looking, quick...

d) Fruits or pulses: albercoques, pumpkins, beans, ...

e) Names that, because they are infrequent, become nicknames: anacletos, doroteos, baldomeros, toribios, cornelios, damianes, casimir os, ramones...

f) Various nouns for food, objects, parts of the body...: sodas, jugs, mines, butts, waistcoats, porrons, fists, eyelashes, bacon, ...

g) Gentilicios: Galicians, mulattos, Catalonians...

h) Compound nouns, formed by two lexemes, most of them achieving an original combination that is not part of the Spanish lexicon, and which obey different structures:

a) verb-
noun: pelagatos, matacristos, pinchapuertas, tragapuertas, faratacarr os, buscavidas, cagatintas, mascaquesos ...

b) noun-adjective: patascortas, perrosgordos,

c) noun-noun: peñalejas

i) New words often achieved by a combination of striking sounds, very frequent the from the letter"ch", a sometimes reduplicated: pachos, peruchetes, chiqueles, chairos, chairos, chiquetos, chuanes, chispes, chismos, chuchetes, chuchos, cachuchas, cachuchines, chichas....A Sometimes this reduplication is of other sounds: tatines, capitotos, patetas, tutos, cucas, pololo...

j) Names of famous bullfighters: arruzas, gaonas, curritos, manoletes...

This is just a small sample of a whole host of the most varied nicknames, which are brushstrokes in the picture in which the life of a village, of any village, is drawn. Generally, it is not a problem for anyone to bear the nickname (I am very proud to be Jarras and Peñaleja), although there are people who do not accept it willingly.

This has led to more than one argument or even to the break-up of relations on more than one occasion. Although it must be understood that not all people can be aware of whether or not each person accepts the nickname that has fallen into their lot or misfortune, because the truth is that, although the insult or personal disqualification is not sought, there are nicknames that are like a slab that hangs over an entire family generation after generation and it is very difficult to get rid of them. For this reason there is a phrase that is used that puts an end to any situation of this type, urging the offended party to put up with the nickname that comes their way:

- Whoever gave you "bodega" should have given you "cámara última".

As we pointed out earlier, it would be very enlightening and suggestive to know the origin of each nickname. Fortunately, we know the name of some who were born in a wood factory, commonly known as La Leva, which in the 50s and 60s employed hundreds of young people and teenagers. When a new worker entered the factory, they were put through the rite of baptism and given what was a nickname, although over time many have become nicknames, passing on to subsequent generations. Some of them denote great ingenuity on the part of the "celebrant":

- "donelías": this nickname was given to a young man who had a small bald spot on his head, like the priests of the time, and as there was a priest in the village called Don Elías, he was baptised like that.

- "picolino": Picoli was the name of a very unattractive comic strip character. When the young man was told that he was going to be baptised with this name, he did not want to and shouted: "No Picoli, no Picoli, no Picoli". And the "celebrant" told him that he had already baptised himself.

- "octopus": this nickname was given to a young man who had very long legs and seemed to get tangled up in them.

- "bajoca": this young man was first given a name he did not like and as he belonged to the Alubias family, he was baptised with this name, which he accepted.

- "novillo", a young man who belonged to the family of the Chirros and because he was very young, he was given this name.

- "The Pope": on one occasion, on the occasion of a religious festival, the parish priest organised a parade and a young man in a white cassock on a float representing the Pontiff, and he has been known by that nickname ever since.

As we can see, the birth of a nickname or nickname is the fruit of ingenuity on most occasions and that is why they are a good example of the idiosyncrasy of the people, of any people.

Looking to the future, it is a reality that the type of life and society in which these nicknames were born is already very different from today's, and that the way our young people, the generations that succeed us, relate to each other is going in a different direction. In spite of this, although some of these nicknames are no longer used by them, it is also true that others are emerging which, as nicknames in principle, serve to baptise one another, although these nicknames will surely have a much shorter life than those nicknames that have been passed down from generation to generation, forming part of the identity of a community and contributing to that particular charm, so evocative, and for some so unknown, of living in a village.

The nickname and its characteristics

1. **Respect: First and foremost**, a nickname should be respectful. It should not offend, humiliate or denigrate the person to whom it refers. A good nickname is one that is given with affection and that the person accepts with pleasure.

2. **Relevance:** It should have some kind of relevance to the person it refers to. It can be related to their name, a physical characteristic, a personality trait, or a story or anecdote.
3. **Originality:** An original nickname can be fun and memorable. Try not to make it a common or generic nickname.
4. **Acceptance:** It is important that the person to whom the nickname is given accepts it and feels comfortable with it. If the person feels offended or uncomfortable, it is best to stop using it.
5. **Positivity:** A good nickname should have a positive connotation. Avoid nicknames that highlight negative aspects or that can be interpreted in an offensive way.

Finding the perfect nickname depends largely on your interests, personality and preferences. Here are some tips to help you find a nickname you like:

Reflect on your interests: Think about your hobbies, favourite activities, sports or whatever you are passionate about. Often, nicknames related to your personal interests can be a good choice.

Consider your personality: Are you outgoing, introverted, funny, serious or creative? Your personality can be a source of inspiration for your nickname. For example, if you are a friendly person, you might consider friendly and nice nicknames.

Play on your physical characteristics: Sometimes nicknames are derived from physical characteristics or distinguishing features. You can take into account your hair colour, height, eye colour, etc., to create a unique nickname.

Use an online nickname generator: There are many online tools that automatically generate nicknames based on your interests and preferences. You can try some of them to get ideas.

Ask friends and family for input: Sometimes people close to you may have creative ideas for a nickname that fits your personality. For example, a fun "nickname for sister" may come up during a family gathering.

Be creative: Don't be afraid to be creative and think outside the box. Unique and original nicknames are often the most memorable.

Try different options: Don't rush into choosing a nickname. Try different options and take the time to decide which one you like best.

Make sure it is appropriate: Make sure your nickname is appropriate and respectful. Avoid using nicknames that may be offensive or inappropriate.

Discursive modes of nicknames

The communicative unit of discourse is complex to define and depends on the perspectives from which it is viewed. As García, 2010 says, "in recent years, discourse has become one of the key words for many disciplines and has been approached from different perspectives...", generating a certain confusion due to the interrelations between the linguistic and social sciences. We consider discourse as the use of a verbal structure, a communicative and cultural instrument, and a form of socio-communicative interaction in a given context and situation. If we consider discourse as language and as a social, linguistic and textual communicative practice, we find different modes: social, administrative, academic, institutional, legal, journalistic, conversational, instructional, convivial, pragmatic, literary, narrative, descriptive, intellectual, religious, political, conservative, progressive, mystical, realistic, fantastic, transcending, and so forth. "light", etc., and popular. We are going to focus, in particular, on rural popular convivial discourses. And within this discursive category, different subcategories will be addressed in order to contemplate the rich range of discourses produced in these environments.

All of these are susceptible to nuance, given the flexible and polysemic nature of the term discourse. In our opinion, we will link it preferably to the principle of appropriateness, to the communicative register in which it is constructed, taking into account the social use, the situation of the sender, the receiver(s), the communicative intentions and all the circumstances surrounding the communicative context in which the discursive communicative action takes place.

And it is in these societies where certain discursive units which are characterised by their capacity for synthesis acquire great relevance, such as the aforementioned nicknames, expressive greetings with interjections such as "hey!"with highly significant tones and prosodic cadences from a relational point of view, since they mark degrees of empathy, trust, distance, complicity, etc.; and others which, by way of example, show the synthetic character of speech in the villages, as is the case of some expressions used by peasants and shepherds to handle and drive cattle: in the in the case of aballerias and mulares, arre 'go forward or walk', so 'stop or stop', güesque 'turn left', güellaó 'turn right' (Mangado, 2007); or some paralinguistic sounds that indicate go forward or stop to the rhythm and degree in which sounds are pronounced, such as chell-chellchell.... 'to move steadily forward', soh-soh-soh... 'go slowly and stop'.

Nicknames in the neighbourhood or communal world

There are anecdotes that are highly representative of the social use of nicknames that are repeated in almost all the neighbourhoods organised into Communes. We can cite the almost impossible identification of some neighbours by their official name, since everyone recognises them exclusively by their nickname. This is the case of Vida suave, Cucaña or Baile en bote, among many others, in the Comunas we have taken as a reference, where some anecdotes are told, such as not being able to identify some neighbours by their first and last names, but always by their nicknames, which is a clear indicator of the great power of these nicknames to identify them precisely. This This phenomenon has led to the creation in some villages of specific "telephone directories" using nicknames, due to the lack of knowledge or confusion as to the official identity of many of them. There is also the case of small businesses that end up being called by the nickname of their owners, as in the specific cases of El Serrano and Canejo. It is also quite common in Asturias and in some villages in La Rioja, among others, where the death notices add the nickname for better identification and as a sign of identity that is widely accepted by the families, as in the case of the aforementioned Capitán, Chospas and Tacuesa. There are even some families who add it to their gravestones. All this gives an idea of the extent to which these appellatives are used in the rural world.

In the field par excellence of the creation and survival of nicknames, we cannot forget the "aesthetic" or creative function of nicknames (Ramírez, 2005a). In almost all villages there are people with a great ability and tendency to assign nicknames to their neighbours due to a special intuition and capacity that allows them to capture attitudinal, corporal, behavioural signs, etc., which leads them to create words as lexical caricatures of the nicknames. As we said before, the poet Federico García Lorca was considered as a great nickname-giver. Perhaps this is the reason why the names of some of the characters in his dramatic works are so relevant and accurate: Yerma, Angustias, etc.

Social use of nicknames

We have already mentioned some of the social uses of nicknames as identifiers and as expressions that generate bonds of trust, positive complicity and familiarity. Most of the above-mentioned referred to accepted and positive nicknames, but we must also consider those which connote negative aspects for their bearers and which imply misunderstandings, misunderstandings and anger. In some cases they are flatly rejected by the nicknames, although this does not mean that they are lost: social reality shows that the worse a nickname is accepted, the more it remains in force and in use, even if it is not used in front of the nicknamed person. As is often said in the villages, the best way to earn a nickname and perpetuate it is to show anger and disagreement with it; automatically, the people tend to register it as a preferential identifying element of the person who rejects it: there is a certain collective complicity in reinforcing this nickname and a point of consensual stubbornness in doing so.

As for the acceptance or rejection of nicknames, there are several reasons why they are assumed or rejected, as well as their relevance depending on the situations, contexts, by whom and in front of whom they are used. It is obvious that some refer to aspects, reasons or anecdotes that are tolerated and even "liked" at certain times, although not in all situations. We have to take into account that some of the nicknames arise in the families themselves (therefore, without offensive intention), others in the school environment, in youth gangs, in sports, leisure, work, professional circles, etc.; and the rest are inherited as a family and social heritage in which, as in many proper names, they have already lost their straight and original semantic value.

What are the most accepted nicknames? We can say that those that do not refer to offensive, rude, disqualifying, insulting meanings, etc., are accepted without any great problem. And the same happens with those that are inherited and their origin and their straight meaning is lost in time. Nicknames such as Tecle, Risio, Sopas, Ajito, among many, many others are accepted, even with pride, by most of the villagers, although in the village, yes. Also because of what they imply in terms of heritage and belonging, as in the case of expressions such as "Yo soy de los Pandos, Casquetones, Carinas, Pichoches, etc.", or the aforementioned cases of their use in companies, obituary notices and even gravestones. An example of the importance of nicknames can be found, for example, in other environments, such as in Elvira Lindo's literary Manolito Gafotas, when El Orejones comes to say that it is better to have that nickname than none at all because, if not, "you are nobody". These are examples and realities that show the value of nicknames as personal signs of identity and family or social belonging. And which are the least accepted or rejected? Obviously, those that refer to disqualifying meanings or circumstances not accepted by the nicknames. It is quite understandable that real nicknames such as Mierda, Basura, Mocazos, Morrotorcido, etc., are unacceptable to the nicknamed, especially because many of them arise from negative evaluations, adverse circumstances, disabilities, physical appearance, painful experiences and inadequate or disqualifying behaviours. Moreover, as discussed above, if the nickname

is not accepted, it tends to become more entrenched. and to be used more frequently and become a source of derision and laughter by those who use them to identify the nickname, even in their absence. On the other hand, it is well known the jocular, sometimes mocking and somewhat malicious character of the "professional" nickname-giver, who is regularly recognised for his intuition, spark, creativity and, as they say in the villages, a little bit of "mala leche".
In this sense, our opinion and attitude is that we must be very respectful in the assignment and use of nicknames, avoiding all those with negative and offensive connotations and which are not accepted by the nicknamed. For this reason, we have carried out educational research-action research in several schools and educational centres in order to reflect on the appropriateness or not of the use of nicknames, especially those that may involve a painful and traumatic aggression for some people, especially in the infant and juvenile stages.
However, we have to say that most of the inhabitants of rural societies over 35 years of age use them as a habitual formula. And it is usually the case that those who use nicknames and nicknames most frequently are, in turn, those who accumulate the most nicknames for their own identification, distinction and characterisation. In any case, and without downplaying the sensitivity of the subject, which it undoubtedly is, in all the research it has been necessary to adopt a certain eclectic attitude, but sensitive and committed to the values of coexistence. It has been very convenient to overcome the prejudices surrounding the issue of nicknames and some dogmatisms that would have impoverished the study and educational actions (Ramírez, 2003 and 2005[a]) and overshadowed the usefulness and the eminently rural heritage that many of these nicknames constitute for those who use them with common sense and good faith. An example of this is the news that appeared in the media about the publication of the nicknames of Huétor Vega, Granada (Pérez-Rejón, 2002)[8] on Tele 5 on 1 June 2001 and in the Ideal de Granada of the same dates, where R. Urrutia writes the following:

The residents of Huétor Vega were so pleased to see their nicknames published in a book that after the first edition of Huétor Vega y sus vecinos, by Francisco Pérez-Rejón Martínez, which consisted of a thousand copies, was sold out, a second edition has been published with another 500 copies. The book contains practically all the nicknames of the residents and families of Huétor Vega over time, many of which still remain with the younger generations. The popular form of the text, the intense research work carried out by the author and the affection with

which all the neighbours and their nicknames are treated, meant that the first edition was sold out in a short time. Almost a third of the copies have been requested by emigrated hueteños, some even to Latin American countries. The book contains more than a hundred nicknames and nicknames from up to the beginning of the century and includes sections dedicated to the mayors, the different priests, the judges or the squares and irrigation ditches.
We believe that this is a good testimony of a positive evaluation of this cast of voices, as well as of the way in which nicknames are collected, catalogued and studied in order to achieve their acceptance. This contrasts with others where misunderstandings, displeasure and even the occasional conflict occur. Pérez-Rejón's approach to nicknames guides us towards the approach and attitudes with which these investigations should be undertaken. And a good example of this is the very positive reception that the article comments on, and which connects with the theme of values, precisely in the affirmation he makes when he speaks of the "affection with which all the neighbours and their nicknames are treated". Surely this good treatment of the subject of nicknames with large doses of professionalism, calm, empathy, common sense and human quality in relationships, among other reasons, is one of the reasons why the study and the book has been so much in demand by the inhabitants or descendants of Huétor Vega.
As a pragmatic discourse, the nickname is a highly profitable and commonly used speech act: we can say that, in the villages, practically all the neighbours have their own nickname with a high degree of identifying and distinguishing capacity. Some of them carry more than one, the familiar one and other personal ones that they have acquired at some stage of their lives for different reasons. The use of these nicknames is daily and natural for the rural population and is also maintained in some urban contexts with rural roots and awareness of them. In villages, the use is usually direct and unambiguous; in cities, it is usually used in specific contexts evoking the peasant life to which one originally belonged.
In both cases, different degrees of respect and empathy are perceived depending on who uses them, to whom they are mentioned, with what intention, at what moment, in front of whom, in what situation and circumstance, in what communicative key, in what climate of trust, complicity and reciprocity. This widespread use of nicknames, then, gives us more than enough evidence to classify them as synthetic discourses of a very practical nature in social relations, especially in rural areas. The fact and the aforementioned reality of having to resort almost obligatorily to the use of nicknames to identify people confirms their

synthetic and pragmatic character: Fernando Martínez there may be several - in the villages names and surnames are often repeated - and may be unidentifiable, even after tracing the surnames and their paternal and maternal ancestry, but Mielero there is only one, Nano Mielero; the same could be said of Jesús Ruiz, who there may also be several, but Forris there is only one, Chuchi Forris; and many other similar cases. The same can be said of the telephone directories by nicknames that have been created in some villages, sometimes municipal, sometimes business or associative, in order to be able to correctly identify neighbours. Likewise, we can cite the habit mentioned of naming some companies by the nickname of the owner or the family (excavators Canejo). But perhaps the most emotive example of the usefulness of nicknames is their use in death notices (Milagros Chospas) and on gravestones in village cemeteries (Ángel Capitán).

CONTEXT: AN APPROACH TO THE COLOMBIAN PACIFIC

The Colombian Pacific Region is one of the richest regions in terms of biodiversity and resources of all kinds. To get an idea, it is enough to mention a few examples: it is estimated that the region has one of the highest rates of continental endemism of plants, that is, species exclusive to a terrestrial region, 8 or 9 thousand species of plants out of the 45 thousand that exist in Colombia; there are also estimates that 11% of all known bird species in the world and 56% of Colombian species are represented in the region. There are 10 million hectares of great ecosystemic variety that allow the region to be recognised for its great ichthyological wealth, with an estimated 400 species of freshwater and marine life, and aquaculture accounts for about 20% of the country's fish production. In terms of minerals, it is known for its large deposits of gold and platinum, other minerals of potential use are copper, manganese, chrome, iron, coal and magnetite. As if that were not enough, ECOPETROL estimates that 36 million barrels of oil and some 45 million cubic metres of gas. The Pacific region also has a potential of 248 rivers, integrated in the Pacific river basin, which are vital for riparian communities and the world where the resource is limited.

The Pacific region is characterised by the existence of strategic ecosystems with immense potential that must be protected. Because of its biodiversity, the Pacific is recognised as one of the most privileged places on the planet and is a strategic point for the country's insertion in the world economy and a fundamental factor for its competitiveness.

Seventy-nine per cent of its ecosystems have not been transformed; the region has four national natural parks and one wildlife sanctuary; the region has been declared a forest reserve for the protection of soil, water and wildlife. However, despite its great potential, the Pacific is an under-studied region: only 1% of researchers and 2% of institutions work in the Pacific.

The other side of the coin in the Pacific is the fact that the region is the poorest in the country according to official sources. All the indicators by which development is measured in the Pacific are the highest. For example, while in Colombia as a whole functional illiteracy (less than three grades) is 15%, in the Pacific this percentage rises to 18%. In relation to health, infant mortality, which is around 19 per thousand in Colombia, is 54 per thousand in the region, malnutrition in Colombia is 13.6%, and in the Pacific this figure is almost double: 24%. The departments of Nariño and Cauca have the highest rates of chronic malnutrition: 24%, while the national average is 13.6%. With respect to the country, the Pacific Region has the highest rate of undernutrition due to low height for age in the 10-17 age range.

Talking about the ethnic peoples of the Pacific, both indigenous and Afro-descendant, means first of all recognising that these peoples are contributing to the current cultural, environmental and social structure of the region, based on their own and imposed dynamics, which are reflected both in their processes of physical and symbolic resistance and in their processes of adaptation and syncretism, which also include the rise and crisis of their identities, as well as their migration processes. It is also necessary to consult the regional history, with its processes of population and mobility, its relations with national and regional society, the waves of colonisation, the extractive booms and the expansion of the market economy system, facts that have undoubtedly generated profound transformations in their territories and cultural systems. According to the National Council for Economic and Social Policy - CONPES 3491 (2007:6), a state policy for the Colombian Pacific is proposed which contains the application to the Pacific of the policy "Community State: development for all". It aims to insert this region into national and international development within the framework of a strategic programme for social and economic reactivation, which seeks to improve the living conditions of its inhabitants and takes into account the region's natural and ethnic ecosystemic conditions.

THE PACIFIC: A LINGUISTIC COMMUNITY

Tumaco is a linguistic community because the vast majority of its inhabitants are active members of "...a group of human beings who use the same language or the same dialect at a given time, which enables them to communicate with each other..." (Jean Dubois, 1979:30). Similarly, a linguistic community is "... a group of people who form a community, a region, a municipality, a neighbourhood, a nation, and who have at least one variety of speech in common..." (Richards et al. 1992).

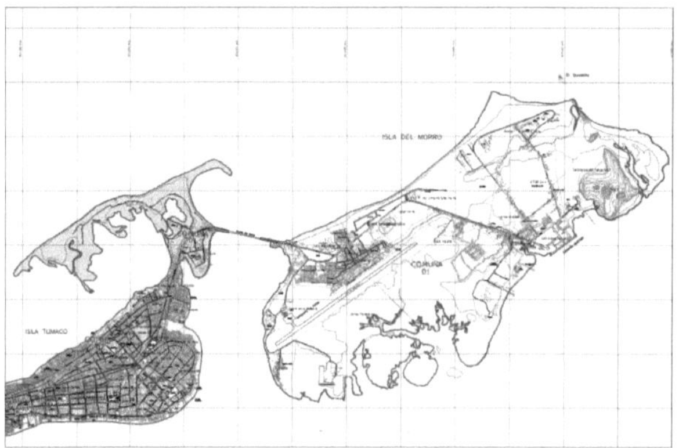

Well, a language community must be defined especially in terms of the varieties it uses, its register and linguistic norms of interaction as well as its skills1. For the above, it is necessary to bear in mind that... "...a linguistic community is not homogeneous; it is always made up of a large number of groups with different linguistic behaviour; the form of language that the members of these groups use tends to reproduce in one way or another, in phonetics, syntax or lexicon, the differences of generation, origin or residence, training, or profession and socio-cultural differences". (Dubois. J. 1979).

In the above context, it must be pointed out that in addition to reciprocal communication between Bonaerenses through a common code, there is also the possibility that some members of this community share two or more codes different from that of the larger community; such is the case of indigenous communities. Thus, a linguistic community can be divided and subdivided into numerous lower communities; an individual can belong to several linguistic communities at the same time. Finally, the concept of a linguistic community only implies that certain specific conditions of communication are met, at a given time, by all the members of a group and only by them; the group may be stable or unstable, permanent or ephemeral. Socially and/or geographically based. The Pacific Region is a linguistic community because most of the people who inhabit it communicate through the Spanish language use the general Pacific Coastal dialect and particularly share the predominant Coastal dialectal variant. Nicknames or nicknames have a universal character and have been used since the beginning of time in all human societies as predecessors of proper names and surnames. They have been and are appellatives used in close circles to accurately identify the people they nickname. It is common to feel them consubstantial to rural societies and, consequently, to forms of speech of a popular and colloquial nature, far removed from the official uses established by the cultured norms of treatment. Sometimes, due to the significance of some of them, they are considered as offensive nicknames and it is not uncommon to find certain resistance to being called that way by many people. Certainly, some nicknames are far from being pleasant and positive words for those who use them, although on other occasions they do refer to more accepted meanings. However, in the case of nicknames, as in almost all facets of life, things are neither black nor white. At least not entirely. Reality turns to show its magnificent complexity. Indeed, they are neither exclusive to villages, nor are they always rejected; thus we find these nicknames not only in rural areas, but also in any circle of proximity: group of friends, schools, neighbourhoods, sports teams, groups of artists, etc. Similarly, nicknames do not always have negative connotations and are not always rejected by the nicknamed. And we also find that the acceptance or not of this informal and unofficial nickname has a lot to do with the degree of trust, emotionality and complicity of those who use it. This article aims to show how nicknames constitute a synthetic and very profitable discourse, due to the economy of language they entail, as well as clarifying and generating coexistence ties and highly creative linguistic productions. The poet Federico García Lorca was himself a great a procurer during his time at the Residencia de Estudiantes in Madrid. And both he and other authors such as Miguel Delibes (El camino, Las Ratas,) and Camilo José Cela (Tobogán de hambrientos), to give a few examples, typified many of their characters by means of nicknames that became highly accurate metaphors and of great identifying precision, like a linguistic caricature of the aforementioned characters in their literary works.This work has its origin and background in the author's interest in the ways of life in rural societies, which led him to develop his doctoral studies, dissertation and doctoral thesis on nicknames, as well as in other research on onomastics that he has been developing in recent years and in various publications on the subject. But the pristine origin lies in the author's belonging and links to the peasant world in which he was born, raised and lives. Hence the concern and occupation of compiling and studying these nicknames, as well as analysing them not only from a linguistic and sociolinguistic perspective, but also from their concrete uses and what they imply as elements of relationship and coexistence. To some extent, over the years, the aim has been to recover, through these nicknames so deeply rooted in rural relations, part of the intangible heritage at risk of disappearing at the same time as the agricultural and livestock world in which they have been most developed is in decline.This paper aims, therefore, to offer and share a brief overview of a subject that we believe to be original and on which we are constantly working.1 It consists of several sections in which the subject will be dealt with from a threefold perspective: 1. 2. The urban neighbourhood society organised in five (5) communes, and some synthetic discourses of coexistence. 3. The social use of nicknames in the urban neighbourhood world.

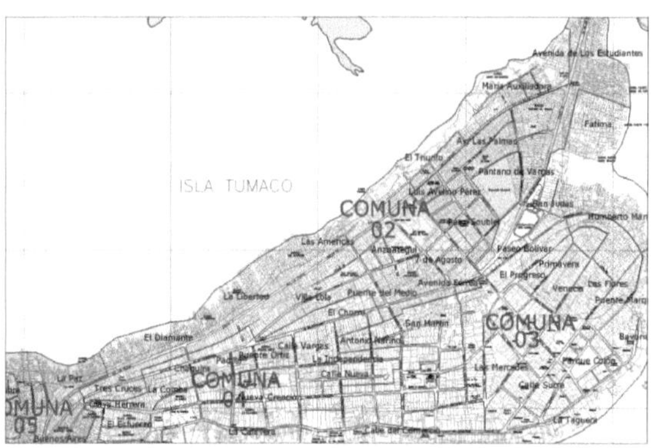

METHODOLOGY

This work is framed within the dialectal type of research, because it describes a plot or aspect of a language: the nickname, in a place and within a social context, Tumaco, with a synchrony in the data collection, at the end of the beginning of 2012-2013, and uses techniques or methods of dialectology, a discipline considered as the branch of linguistics, which "must give reason for the variety and intradiasystematic variation" (Montes, 1995, p. 115). 115), i.e. the reason for the variation or variations of a language in a geographical space or, as Luigi Heilman, quoted by Montes, put it, "the study of unity in variety, i.e. the way in which a set of norms, varieties and variants are integrated into a larger whole" (Montes, 1995, p. 71). Now, if a language is a system of signs produced by a community or society in its inter-individual communication - a social fact - and dialectology, with its method, linguistic geography, makes it possible to collect part or all of this form of linguistic expression, produced by speakers in a community, determined in time and limited spatially and socially, this research conforms to the linguistic patterns of the dialectal discipline.

DATA COLLECTION

The collection of information is the practical period of the research process, the fundamental part of it. It allows, from the fieldwork, to establish the state of the question or phenomenon under investigation. Linguistic research in Colombia has recently experienced a great boom, thanks to the impetus that the Caro y Cuervo Institute has given to the study of Colombian Spanish, together with the efforts and work of institutions and centres of higher education, which have assumed with responsibility and commitment the need to study the linguistic system that we use as a means of communication. Some brief or extensive work and research on this topic has been carried out so far. However, much remains to be done. This is, therefore, a further contribution to the knowledge of Tumaqueño Spanish. The following were taken into account for its realisation:

The questionnaire

A short questionnaire of 19 questions was elaborated and applied, divided into two main groups, as follows: A- Generalities and B- The phenomenon to be investigated. The first part corresponds to: name of the informant, place of birth, sex, age, profession or trade, and the sector where the informant lives. The second part has to do with the nickname itself: nickname, origin, motivation, description, physical description of the nickname (characteristics), time of the nickname, who gave him/her the nickname, where it is given, why it is given, likes it, why he/she likes it, does not like it, why he/she does not like it, and some observations. From the methodological point of view, the questionnaire is one of the most frequent resources, and being useful, important and practical, it has been used since ancient times in linguistic research, "because it concentrates the object of study in an unsurpassable way, it directly addresses the subject with unlimited power of expansion and depth and, as expected, it produces linguistic data with accuracy and economy" (López Morales, 1994).

The survey: the surveys were carried out during the last months of 2012 and the first months of 2013. The survey work is the result of the work done by me, as a language teacher, tutor in the area of language and mathematics in the PTA and PNLE programme of the Ministry of National Education, and as a member of the Redlenguaje de Colombia, in accompanying teachers in reading and writing. There I had the opportunity to guide final works of teachers who were developing classroom projects, in relation to the collection of lexicographic information and dialectological desinences of neighbourhoods and villages in the municipality of Tumaco. As a task of revision and verification of the work carried out by teachers and students in these academic programmes, I carried out several complementary surveys.

At the beginning, my intention was not very ambitious and consisted of simply collecting a small sample, in order to present a few observations on the phenomenon of nicknames in the bulletin of the Pacific Language Node of the Colombian Language Network. However, with those first collections and a talk with Linguistics professors, I realised the importance of expanding the sample for the work of the degree. I then carried out a final questionnaire, which allowed me, in the company of the students, to collect this very rich material which constitutes the linguistic corpus of the work.

Thus, the first part of the surveys could be called exploratory work on the linguistic reality to be investigated. The second part, the actual work itself, was the collection of the corpus, in which no less than 96 students and 40 teachers participated. The surveys were only intended to collect sufficient material for this study. My interest with the students was to induce them to practice and collect a sample of a particular speech phenomenon in the population of Tumaco and its incidence in the Comunas: The nickname. This task was carried out and evaluated as final course work and with the teachers as support for the articles they were writing about the phenomenon. Hence, the surveys were of varying lengths. The initial, or baseline, surveys were carried out in the second half of 2012. These allowed me to explore the linguistic phenomenon and, subsequently, to establish the validity and reliability of the final instrument to be used. The other, definitive surveys were carried out in the first and part of the second half of 2013.

The questionnaire was tested in a pilot survey, applied to some colleagues from the Language Node and to teachers from the institution where I work, men and women I know, which allowed its validation and reliability.The final instrument was applied, by final grade students studying strategically in the five communes, to men, women, young people and adults, native Tumaqueños, residents, immigrants, from different professions or trades, and living in Comuna 1, 2, 3, 4 and 5 of the municipality of Tumaco.

Due to the large number of students in the last grades of the Media Académica, it was possible to collect a lot of material. For this purpose, the students were divided into groups of 5 and each of them had to apply at least 3 surveys, among family members, friends, acquaintances, colleagues, neighbours, etc., i.e. each group brought me 15 questionnaires. This is why the sample was so large and very rich.

The informants. The informants are the product of the selection that the students determined, according to the given parameters. That is, according to the degree of friendship, familiarity, neighbourhood, etc., the interviewer applied the questionnaire. At no time was it required to apply the questionnaire to a specific group or number of people, nor were they defined by gender, age, or origin. Hence, the informants constitute a heterogeneous group of men and women, young people, adults, professionals or not, foreigners or natives, living in the different neighbourhoods and sectors of the capital city.

The sample. It is made up of the 750 records collected during 2012 and 2013, in the surveys carried out through the final questionnaire, for a total of 1,250 nicknames.

ANALYSIS AND INTERPRETATION OF THE RESULTS

The following graphs correspond to some of the questions in the questionnaire, selected to clarify aspects of the function of apoco in Tumaqueño speech.

Graph 1 Question N°6 Time of nickname.

GRUPO I	47%	MENOS DE 5 AÑOS
GRUPO II	14%	ENTRE 5 Y 25 AÑOS
GRUPO III	31%	ENTRE 5 y 15 AÑOS
GRUPO IV	8%	MÁS DE 26 AÑOS

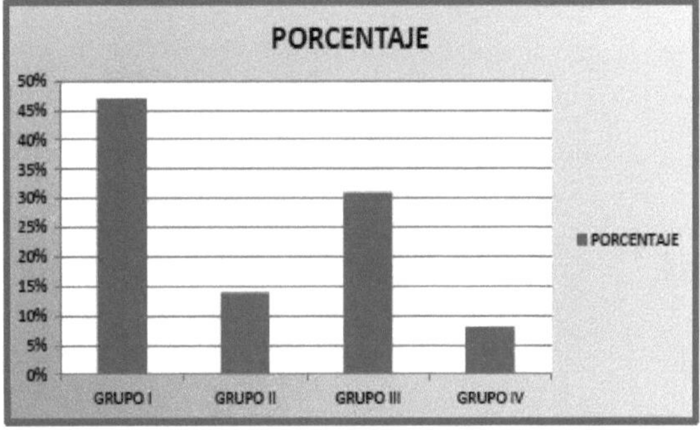

The graphic visualisation of this question allows us to corroborate that the nickname, as a linguistic phenomenon, marks the subject from childhood until one ceases to exist. See Thus, four major groups can be observed, indicative of the time of the nickname, ranging from early childhood to maturity and certainly beyond the death of the person possessing the unfortunate or fortunate, as the case may be, speech act.

Figure 2 Question N°7 Who gave you the nickname?

PERCENTAGE

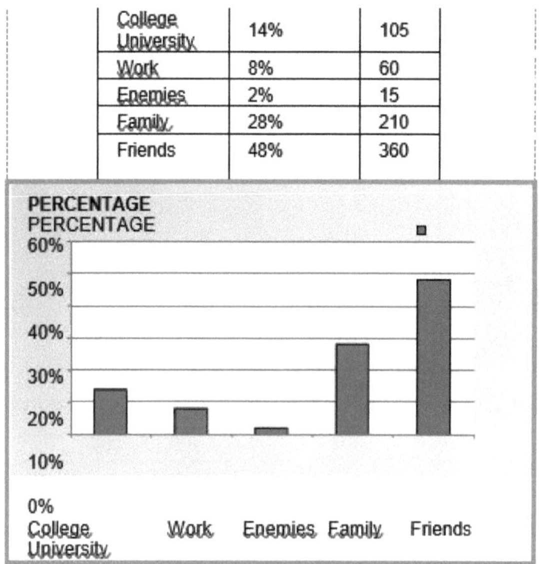

College University	14%	105
Work	8%	60
Enemies	2%	15
Family	28%	210
Friends	48%	360

When asked about the author of the nickname by which they were baptised, 48% of **informants** answered friends; 28% family members; 14% school or university; 8% work colleagues; 2% more than enemies, and a very low percentage who do not know it. This shows that it is indeed the "friends" who are the ones who give the most nicknames and who, for various reasons, achieve their goal. They are the ones with whom most time is spent on a daily basis. Relatives do it out of affection. Enemies do it for personal reasons. At school or university, classmates enjoy and share with the subject the fate or misfortune of the nickname. If we join the latter to the group of friends, we conclude that those who are closest to the subject sentimentally or emotionally are the ones who mortify him or reward him with a nickname (the friends).

Figure 3 Question N°8 Where do they call you by your nickname?

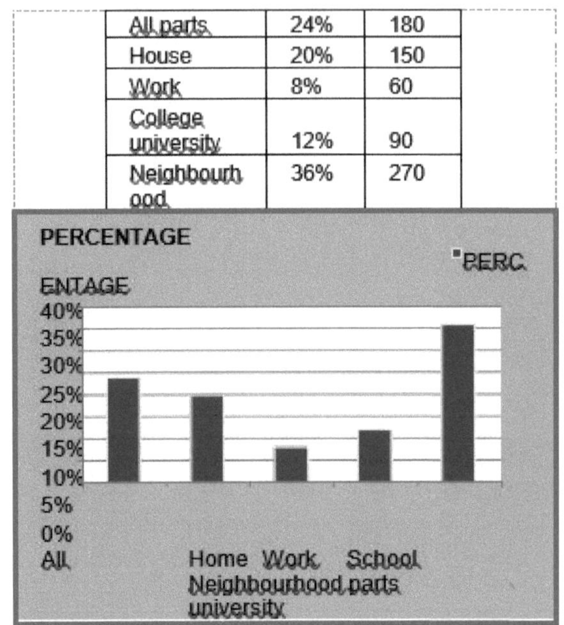

All parts	24%	180
House	20%	150
Work	8%	60
College university	12%	90
Neighbourh ood	36%	270

According to the graph, nicknames are said everywhere, indirectly or directly, to the nicknamed. But it is in the neighbourhood where they are most frequently used. This shows, as in the previous graph, that it is childhood or social friends who wear the mask of friendship to say this or that nickname. In the same way At university, school or work, the companionship allows the indiscriminate use of it. At home the situation is less so, but wherever the subject is, his or her nickname functions as an evocative element of the situation that created it. In short, the nickname functions as a chain, which is born in any place, time or circumstance, and which moves with the subject in all places.

Graph 4 Question N°10 Do you like the nickname?

Indifferent	13%	97,5
No	21%	157,5
Yes	66%	495

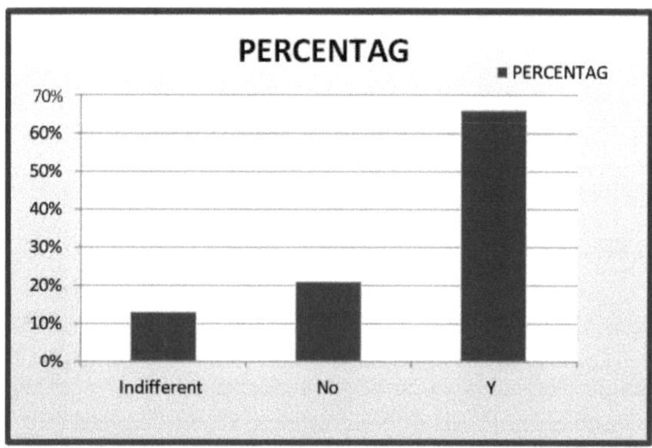

From the graph it can be inferred that the informants (men and women), in a good number (66%), like the nickname; only 21% do not like it, and 13% are indifferent to it. It can be seen, then, that the Tumaqueño speakers enjoy their nickname, are happy with it, or have become accustomed to it. This surely allows them to use it without restriction in everyday life. Of course, the vast majority of the nicknames in the sample are either mild or kind to the subjects, who wear them, as there are also cruel or sarcastic ones, who humiliate or ridicule

Figure 6 Question N°13 Relationship to the interviewer

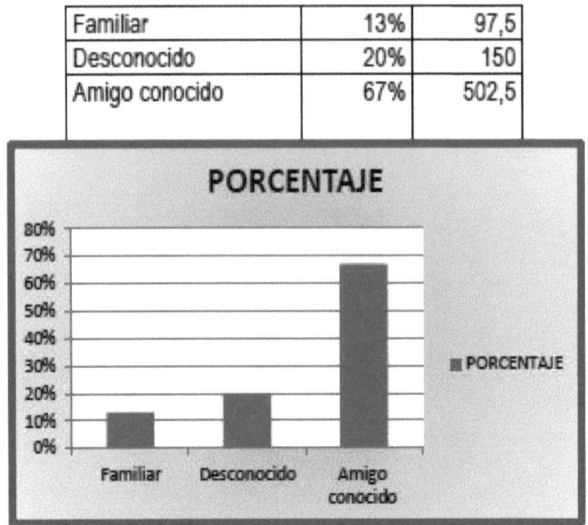

Familiar	13%	97,5
Desconocido	20%	150
Amigo conocido	67%	502,5

The analysis of the graph shows that 67% of the interviewers used their friends or acquaintances (school, neighbourhood, work), 20% used strangers and 13% used the family nucleus. Their knowledge and relationship with the informants made it a little easier to obtain the information. This shows, once again, that contact with people within the social group makes it easier to get to know them or at least to reach the members more easily in order to obtain this type of information.

Figure 7 Question N°14 Sector where respondents live

Commune N° 1	10%	75
Commune N° 2	10%	75
Commune N° 3	20%	150
Commune N° 4	26%	195
Commune N° 5	34%	255

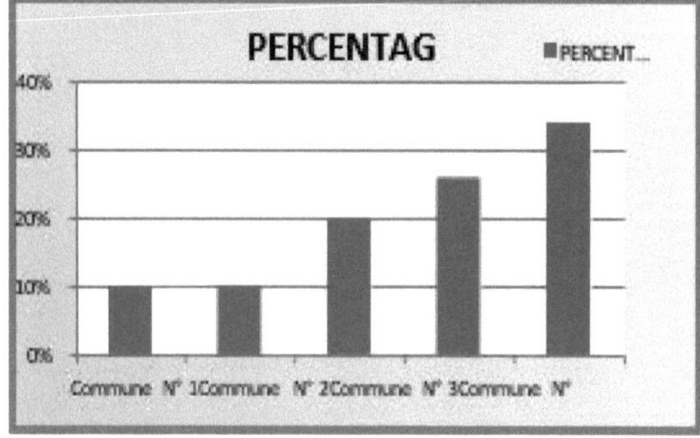

The graph shows that the respondents live in the different cardinal points of the capital city of the municipality. The vast majority of them live in Commune 5 (34%); 26% in Commune 4; 20% in Commune 3; 10% in Commune 2; and 10% in Commune 1. The latter is the group that contributed the least information to the sample. In Commune 2, even though it is a commercial and complex sector, the same number was surveyed as in Commune 1.

Graph 8 Question N°15 Age

Group I (12 to 33 years)	86%	90%
Group II (34 to 57 years)	13%	4%
GroupIII (58 and over)	1%	6%

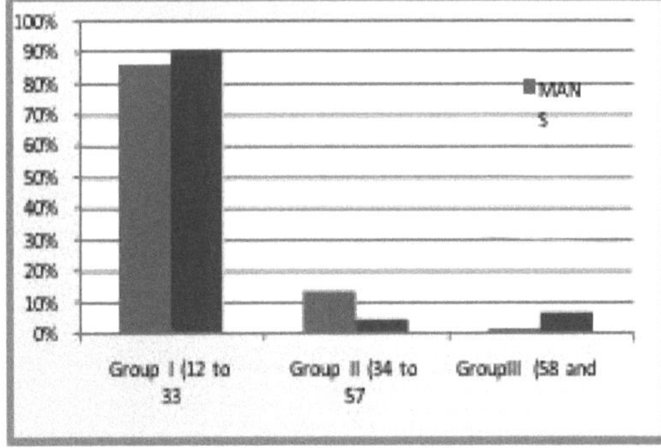

The informants in this large sample are men and women, aged between 15 and 80, divided into three generational groups, as follows:

Group I (12 to 33 years): Males 86% - Males 86% - Males Women 90%
Group II (34 to 57 years old): Men13% - Women 4% - Men13% - Women 4%.
GroupIII (58 and over): Men 1% - Women 6%.

CONCLUDING REMARKS

The nickname is a motivated linguistic sign, very expressive, and as such has connotative meanings, as it communicates only what is referred to in the utterance. It expresses feelings, attitudes, moods of the speaker in relation to the referent or the listener, coupled with the circumstances in which it is produced. It is tinged with a certain emotional, festive, humorous, sarcastic or rude colouring that the speaker who creates or recreates the sign gives it. In its function, the nickname as a linguistic sign (lexeme) leaves aside its denotative sense or objective factor, to give way to the relevant and contracting semantic features that evoke the most stable, outstanding and generic characteristics of the object, subject or reality being compared. Thus, the nickname encompasses or conceptualises real or supposed characteristics of the referent, qualities of the objects and actions, everyday life, cultural, ideological, social phenomena, etc. This makes it current, useful, constant and, in many cases, perpetual, across generations.Through its use and acceptance it wears out, it loses its function of exciting and delighting, it erases what Le Gern called "the associated image", it de-anticises, or as Ullmann (1965, p. 156) said when referring to the loss of the emotive meaning of words, it suffers "the law of diminishing returns"; it loses or disappears from it that connotative sense and begins only to denote, converted into a common and ordinary sign, without any restriction incorporated into the use as a sign of identification of x or y subject. But when this happens, the person who has given the subject a new nickname, an expressive term which, with astonishing skill, he surely recreates or creates again, as the case may be; for when the nickname "gets old" or the person who has given it the nickname feels that it has no real meaning, he creates a new one with which he satisfies his expressive need.In various speech situations, the user externalises through the nickname the burden of sociological repression that he or she harbours, suffers or feels for others. Its use, on the part of the speaker, has a purpose: to provoke a certain reaction, positive or negative, in the person nicknamed, and, in him, an outlet for his passions. As Father Félix Restrepo said in El alma de las palabras (1974, p. 26), one succeeds in "violating the meaning of words so that they mean the opposite of what they say [...], in them one can observe resentment, compressed anger, indignation, disdain or sarcasm.The speaker, in many cases, in order not to offend susceptibilities or simply not to say the nickname directly, uses euphemism as a softening tool or uses detours, circumlocutions, periphrases, until he reaches the mind of others with images that shape or give shape to the nickname (it is much more complex, requires greater concentration and mental speed, but the user is very skilful), thereby avoiding crude, vulgar or tasteless words, with which he probably hurts or satirizes. The intention here is not sarcasm but humour, hilarity; direct or objective reference is therefore avoided.The direct expression of the nickname, to the nicknamed, is avoided in front of him, which allows the nickname to have that feeling of humour and complicity, a feeling that makes it effective in its intention and survival within the group. The speaker hides in some way to express the nickname, because the important thing for him is to achieve his goal and make fun of others. In accordance with the above, the nickname, in its use, reflects the feeling of consideration of the social group (whether it is respect, fear, prudence or delicacy for the other), since this or that nickname is almost never said in front of the subject, except in family, colloquial or violent situations, which in some cases unleashes the anger of the nicknamed person and its use ends in tragedy. For this reason, the nickname is carried until sudden or natural death, that is to say, it lives and dies with the nickname, and even more, after death he is remembered by his nickname. The nicknamed, unfortunately, is the last to know that he has a nickname, and from then on he suffers eternally, condemned to ridicule or being the laughing stock of all.Analysis of the sample indicates that nickname use has degrees of preference. They range from those that are expressed most frequently, to those that are used less frequently, minimally or not at all. Some are rude, denigrating, satirical, vulgar, crude and insulting; others are mild, noble, expressive, jocular or humorous. The latter are accepted, in large numbers, by the speakers and by the unfortunate possessors, because they nestle in them a feeling of affection or friendship, in spite of the defect, quality, vice, etc., which is emphasised in the subject. In both cases, they depend on the person expressing them, their tone and the situation in which they are used. The nickname is either noble or coarse, depending on the speaker and his intention.The use of the nickname is conditioned by the group, the social environment, the circumstances, the intention or desire of the speaker(s). However, according to the contexts of use and its socially elaborated nuance of meaning, it is well or badly received, even if it is widely accepted by the group and especially by the nicknamed, as it is a sign with certain restrictions. These signs are born unexpectedly (when you least expect them) and begin their long or short journey, delighting some and confining others. Some are luckier than others, survive through the ages and are passed on from generation to generation. Others are born and, just as they are born, they disappear. Some are very restrictive and others are widespread. Nicknames, like words, acquire a certain prestige within the group that is transferred to their possessor. In many cases, the nickname is accepted by the nicknamed. He/she does not reject it, feels important within the social group to which he/she belongs, likes the nickname, uses it, boasts about it and expects or wants it to be said to him/her (implicit in it is his/her good or bad reputation, this fills him/her with pride). Of course, this is not always the case: others, on the contrary, dislike it, because they consider it an affront or an affront, they are always prepared for this situation, which unfortunately must arise. The nickname, then, functions as an incentive for pride, mockery or sarcasm of the one (s)he bears it. In both cases, it comes to displace or forget the name and the subject is known more by his or her nickname than by his or her own name.This leads the speaker to annoying misunderstandings, because, as the name of the subject is not known, the naïve speaker uses the nickname believing it to be the real name, creating a real mess. The speaker does not know this or that nickname and, faced with the nickname, suddenly names the object or referent of the nickname, which is intricate in unexpected situations, but which, due to the unusual and naive nature of the fact, fortunately produces laughter and is accepted. The nickname in this way is very funny, because there is no intention to humiliate, hurt or offend; on the contrary, it is the common or ordinary use of a linguistic sign. The "pruritus" of nicknaming is not new, it has existed since time immemorial and has occurred in all peoples and different cultures (Macedonian, Babylonian, Assyrian, Roman, Middle Ages, Renaissance, in kings, monarchs, saints), in modern times and especially in our context, from the most humble to the most pompous: the president, high ecclesiastical or civil dignities, artists, prostitutes, criminals, workers in their different professions or trades, unemployed, etc., are given a nickname. There is really nothing new in it. They are formed now in the same way as in former times. Only the circumstances, the objects, the institutions and the intentions of the creator, which are changeable and unpredictable, have changed. We can say, in summary, that the nicknames of Tumaco correspond to situations and real, concrete or abstract facts of the everyday, city, community or neighbourhood life of its people; it is the reflection or worldview of the objective or subjective world that the speaker of Bogota expresses in this linguistic sign. Not even the person who has been given a nickname escapes from the nominating act, because there are nicknames for him too, they are called: bishop, priest, priest, because with their creation they baptise others (they receive ecclesiastical names that have to do with these things). Nicknames that are not given to the person nicknamed (who has suffered the anguish of this act of speech) because surely the latter, offended by the former, would call his offender: viper, snake, son of a bitch, dog, guillotine, vampire, knave, bat, etc. Experience shows that the nickname-giver is happy to nickname, but is unhappy when it is his turn to receive one, feels stung in his self-love by a hideous snake with larger and sharper fangs, is annoyed, perhaps more so, for he feels anguish and anguish, and is unhappy when it is his turn to receive one, for he feels the anguish and dismay of becoming the laughing stock of all those with whom he once eagerly and happily shared the unhappiness of another or others, thanks to this, his unpleasant **"shrew" or "mojiganga",** which contains all the irreverent perversion and unhealthy passion of the human spirit. In the creation and marking of the nickname there is a kind of game of chance, which favours a few and disfavours many others. The favoured one carries it with nobility, pride, haughtiness, pronounces it before another or others, known or unknown, likes to be called by others, seeks to be recognised by all, is used as a greeting or formula of treatment, is formal and informal; while the other suffers eternally the humiliation, cruelty and irony of an unfortunate linguistic act, carried out, surely, by one of his "best friends" or relatives". While some enjoy, others suffer, but the nickname keeps its function alive.People or parents who know, know or are linguistically aware of the affectation, mark or "stigma" of the nickname on the nicknamed person, try to avoid by all means that the child or son or daughter receives a nickname from someone (relative, friend, playmate at school, college, neighbourhood, street, street corners) that marks him or her for life. If this happens, the child is harshly reprimanded to prevent the nickname from taking hold of his small and fragile humanity, but it happens that, due to malice, wickedness and that which is "forbidden is the most desirable", the nickname begins its mad race and prevails, defeating the kind and just intention of that parent who at all costs wanted to prevent his son from being nicknamed. The nickname, after being placed, overcomes any barrier that stands in the way of its free flight to maximum expressiveness, in the mortal world. There are nicknames applied to the subjects in their youth and bachelorhood, and after marriage to their children, propagating through generations.Some people are happy to have that nickname (to have a nickname), because it characterises them, identifies them, differentiates them, etc., and they do not find it unpleasant like others out there. That is to say, they thank heaven that they have received this nickname, it is better that way, and not in any other way; they accept it kindly. Thus, in each social group, the nickname functions as a synonym of respect, social distinction, in the same name. There are also families or family groups marked by a nickname that identifies or characterises them as a group and individually. In the same way, there are nicknames that do not correspond to the motive that presumably created them, nor do they correspond to the characteristics of the nicknamed subject; they are the result of the strangest, most exotic and implausible occurrences of the communicative act (there is no relationship between subject and object, since the choice of the term is secondary to the intention: to place a nickname), which mark people or families for life. There are nicknames that through constant use become commonplace elements of effective or everyday speech, lose their initial colouring or intention, pass from a restrictive medium to common usage, lose their evocative capacity, becoming an indispensable element of our vocabulary. Hence, many nicknames pass from generation to generation and are used today without the evocative sentiment that produced them. They become, in other cases, identifying elements of high social status, which make the bearers proud.The same subject can have more than one nickname (when this happens one of the two loses ground in use and ends up defeated), and the same nickname can have two or more characteristics. In each of the nicknames there is a set of minimal notional features that the nickname-giver, by virtue of his or her capacity for thought, has managed to abstract from reality. For this reason, it is obvious that in many cases a nickname does not have a single meaning, since it has, from its creation and motivation, several meanings. The speaker skilfully selects, from among the various characteristics of the referent, the elements which are common to both the subject and the object, and from this relationship the nominating act emerges. The lexical creation of the nickname does not occur ex nihilo: it obeys communicative needs, determined by the knowledge of comparative reality, accompanied by rules, moulds or schemes that determine and make possible the act of linguistic creation. For in many cases, these are born within the group as an element of mockery or chat, or as a factual function of communicative discourse. Hence, they are constantly being created and recreated. However, within the dynamics of linguistic signs, nicknames are constantly enriched and loaded with new meanings through use, thanks to the fact that speakers integrate new cognitive experiences into them. The presence of the nickname evokes, in the person's mind, the image of what the nickname represents. The medium, or the world itself, serves the speaker for acts of recreation in which feelings intervene for the qualitative transformation of the environment. Nicknames constitute an open lexical inventory, sometimes very difficult to classify or synthesise; for this reason, several classifications were presented here (by their semantic load, motivating features or semas, the intention of the nickname-giver, grammatical or semantic mechanisms), which come as close as possible to the process of creation in Tumaqueño speech. For a nickname to be recognised as such, it needs the knowledge, the experiences, the values of the subjects and the environment to which it refers, since in its use, it admits innumerable purposes or points of view, whatever the speaker has in mind. It evokes the listener's experiences and knowledge in relation to the referent (emotional, cognitive, ideological, etc.).In the process of creation, the use of the system's own normative mechanisms, common to the formation of words in the language, can be observed. Thus, we find phonetic, grammatical, lexical and semantic processes. In phonetics, there is imitation of sound as an "echo of meaning". Grammatically, in addition to the normative procedures, the speaker, in use, departs from the norm and alters the system in cases such as: gallino, culebrero, tigra, etc. In semantic terms, metaphor and metonymy are used. He also combines procedures combining derivation or composition with metaphor. These types of procedures cover all the nicknames in the sample: onomatopoeias, derivatives, compounds and figurative expressions.However, the most common and important source of creation, according to the sample, are metaphor and metonymy. Comparison is the simplest, most productive and effective resource in the creative act of the nickname. By means of this resource, the linguistic sign acquires new meanings, names new referents. In short, the apodiser by ingenious skill compares two realities and establishes a name by similarity, resemblance or contiguity. The diminutive nickname reflects the intention of diminishing the significant objectivity. Its value is ironic, but also softening and accepted in the social relationship of the speaking subjects. It does not indicate littleness but "linguistic diminution", it is a palliative with which a friendly, affectionate or familiar treatment is achieved. The subjectivity of the speaker and the context in which it is used play a fundamental role. A diminutive nickname can, in some cases, express affection, but in others contempt (contrary to the nature of the diminutive), depending on the communicative intention of the speaker.Intonation, among semantic markers, plays an important role in the context of nickname use. Intonation is the bearer of the speaker's communicative intentions; that is, it softens or gives greater illocutionary force to the meaning, making it acceptable or unacceptable to the group or to the bearer agent. The hypocoristic, used as a nickname, falls into the category of diminutives or emotive expressions that indicate affection or affection. Here the affective predominates over the conceptual, it is really an affectionate abbreviation or simplification of the name. There are hundreds of nicknames in Tumaqueño speech, some of them of a city character that correspond to speakers and realities of the city itself; others, of dialectal origin or product of regional speech, to which a large number of

Tumaqueño speakers belong by origin. Of the sample collected, many are characteristic and are found in other municipalities of the Pacific region of Nariño and, of course, in other parts of the department of Nariño, Colombia and Latin America. It goes without saying, then, without fear of misunderstanding, that there is no place in Tumaco or outside this municipality, village, town, township, inspection, hamlet, cove, river, social group, commune, neighbourhood, large or small, in which its inhabitants do not have a nickname. The opposite is to have it but not to know it, until someone thinks of saying it in front of you. Moreover, the human group, by the fact of living in community, in neighbourhood and according to their profession or trade, occupation or not, characteristics, etc., has for their fellows nicknames that range from the mildest to the most hurtful or vulgar. Some nicknames reflect the "fashion" in all areas or are nourished by it, and others show in the nickname-giver a personal tinge, characteristic of this type of creation. The speaker, through the nickname, tints, paints or colours the linguistic signs, turning them into "effective weapons" to highlight qualities, defects, produce mockery, hilarity, even sarcasm, ridicule, humiliation, express feelings or excite emotions, attitudes (affection, love, hate, etc.), a subtle form of violence through the use of the nickname.), a subtle form of violence through words, with which the unfortunate subject assigned with this or that linguistic act of this nature is praised, beaten, mistreated, wounded or even prostrated on cold stone.Finally, one notes that, in a good number of the nicknames in this sample, there is no tact, delicacy or genius. They are the product of spontaneity and occasion, circumstance or the jocularity of the speaker: their intention is to have fun; a nickname, thus, inspires one and another, with which the group and its possessor happily and amiably enjoy themselves amidst the "collective humour" and the prevailing friendship. Example: a young man who loses a tooth in a fight is called "broken bridge", "toothless", and from his misfortune comes one and another nickname after another ("risa loca", "risa linda", "puente largo", "fin de semana"), of course, in the end only one of them takes hold of the humanity of the subject who is the architect of the festive situation. The success of the nickname lies in the use, the moment and the concrete situation for it.It is clear, then, that the procurer does not create out of nothing; his merit lies in the quick and timely handling of the wealth of themes or images he has in his mental reservoir. Originality, ingenuity and motive are the basis and the success of creation.Topics such as politics (characters: "caneca"), events (shocks, embarrassing situations, humorous), everyday events (love, heartbreak, spite, work, routine, antipathy, being good or bad, rude, stingy, bad tempered, liked or disliked by others, being kind, sentimental, etc.), dialogue situations, use of discourse (crutches, words, turns of phrase, expressions, innovations, etc.), personal situation (how he is, how he acts, how he reacts, what his behaviour is, how he dresses, beliefs), his profession or trade (performance in it, social and cultural position, etc.), defects, vices or qualities, are the most common themes in the creation of the nickname. They enlighten the nickname-giver for his or her nomination. The themes are thus very varied, dissimilar and implausible; however, many of the nicknames that seem original and natural are simply variants of the many integrating themes of the common repertoire of Tumaqueño speakers. These themes, everyday situations and circumstances are translated into nicknames. In short, the "humorous" or "sarcastic" intention favours the creation and application of nicknames in the people of Tumaco. It is evident how the people of the different socio-economic strata of the Communes in which the city of Tumaco is territorially configured use certain words that are framed before society as a marginal language, that is to say a language that is born in the marginal parts. In view of this, it is only logical to think that the city-dweller from Tumaco is in constant contact with other people who use a marginal language that in many discursive occasions its use is not correct, however, its ritual use and the convention that is created around the nickname is totally accepted.

Thus, it is in Tumaco where no matter the space or the situation, a fine sample of the phenomenon we are facing is evident in all kinds of speech acts.

- At the beginning of this research we considered it to be a problem of education, but this was discarded when we realised that it is not only people from the lower and middle social strata but also others from the upper social strata who are using and practising certain neologisms characteristic of marginal languages.
- On many occasions, people from the upper social strata engage in what may be an attempt to create their own marginal and characteristic language; in their use, it is noticeable that they try to create neologisms which in turn possess something of marginality, of isolation, but which have their own characteristics of their world and their spaces.
- It can be thought that it is through this contact that is established in the street, the cuchos (spatially reduced places), supermarkets, street corners, shopping centres, mass media, etc., where people are exchanging their words, that is, there is a network of contact where the interlocutor community has a close link with the new forms of nominalisation (nicknames) that are evident in the city, the speakers independent of their conditions, strata, levels of education, not only establish this relationship but also

They also appropriate this knowledge and implement it in their daily lives. Something that makes other speakers are implementing these forms of naming.

- In establishing these forms of naming, it is evident that each of the social strata implements words similar to those that were originally learned. This is what determines that in many cases the social strata are characterised as we have shown here. That is to say, the boundaries of the social strata among the speakers in the communes become blurred when the use of language becomes simultaneous and accepted among them regardless of the standards of living, the ways of life, and the family and social customs they often experience.
- The speakers of the city of Tumaco are a functional part of avant-garde media such as the internet. This medium constitutes another meeting point where everything discursively converges, since most of them, with greater participation of young people, today are in almost continuous contact with a computer and it is the nature of the internet to erase the here and now of communication, by which they have the facility to make everything linguistic (speech) available to everyone from its formation to its understanding.
- Not only is the internet one of the newest existing sources of contact; it is also possible to talk about television, from which it is also possible for all the inhabitants of Tumaqueños to interact with the world and other communities, whether close or distant, where other types of strategies are implemented such as listening and memorising, as we see how on many occasions words become fashionable thanks to a protagonist of a television series.

ANNEXES

QUALITATIVE AND QUANTITATIVE ANALYSIS AND INTERPRETATION OF SAMPLES

Graph 1 Question N°6 Time of nickname.

GROUP PERCENTAGE

GROUP I	47 %	LESS THAN 5 YEARS
GROUP II	14 %	BETWEEN 5 AND 25 YEARS
GROUP III	31 %	BETWEEN 5 and 15 YEARS
GROUP IV	8%	OVER 26 YEARS

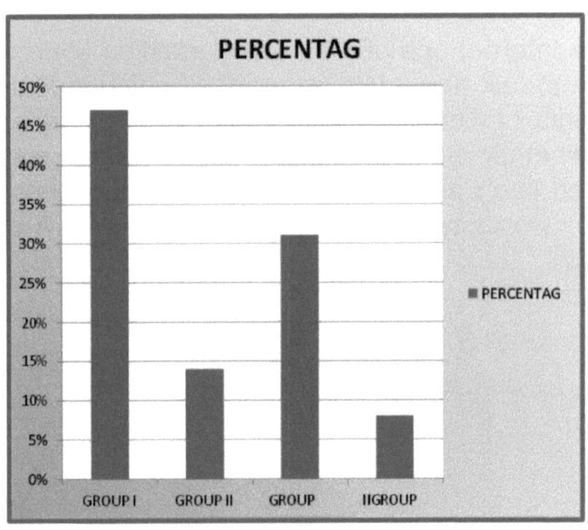

Figure 2 Question N°7 Who gave you the nickname?

PERCENTAGE		
University College	14%	105
Work	8%	60
Enemies	2%	15
Families	28%	210
Friends	48%	360

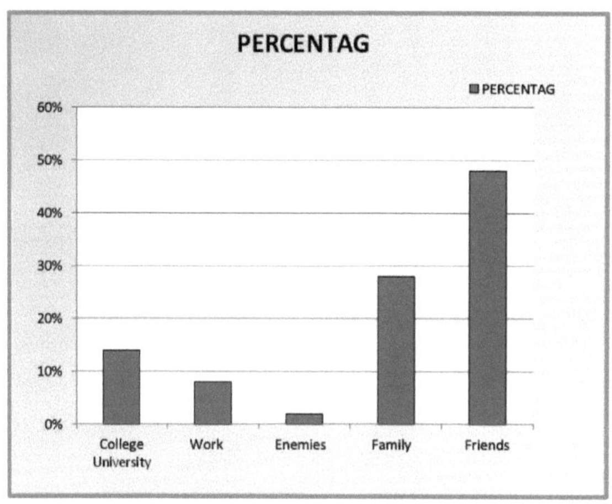

Figure 3 Question N°8 Where do they call you by your nickname?

PERCENTAGE		
All parts	24%	180
House	20%	150
Work	8%	60
College/University	12%	90
Neighbourhood	36%	270

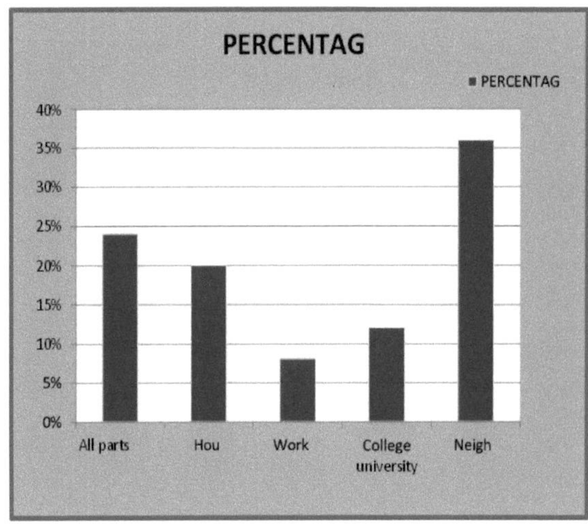

Graph 4 Question N°10 Do you like the nickname?

PERCENTAGE

Indifferent	13%	97,5
No	21%	232,5
YES	66%	495

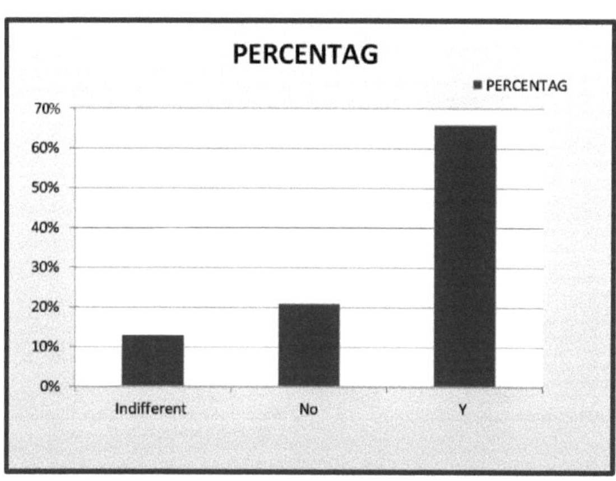

Figure 6 Question N°13 Relationship to the interviewer

PERCENTAGE

Family	13%	97,5
Unknown	20%	150
Known friend	67%	502,5
	100%	750%

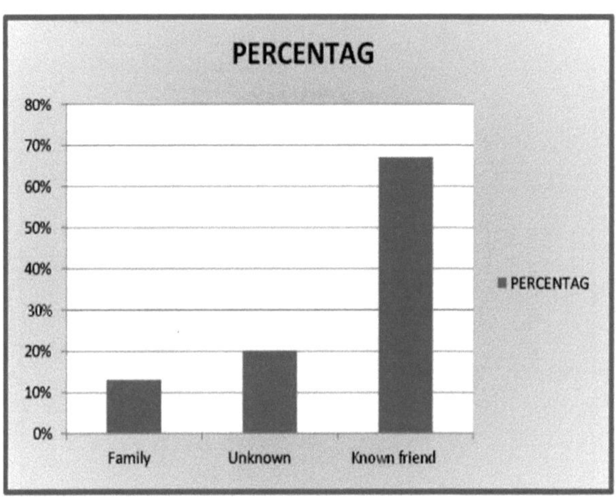

Figure 7 Question N°14 Sector where respondents live

PERCENTAGE		
Commune N° 1	10%	75
Commune N° 2	10%	75
Commune N° 3	20%	150
Commune N° 4	26%	195
Commune N° 5	34%	255
	100%	750%

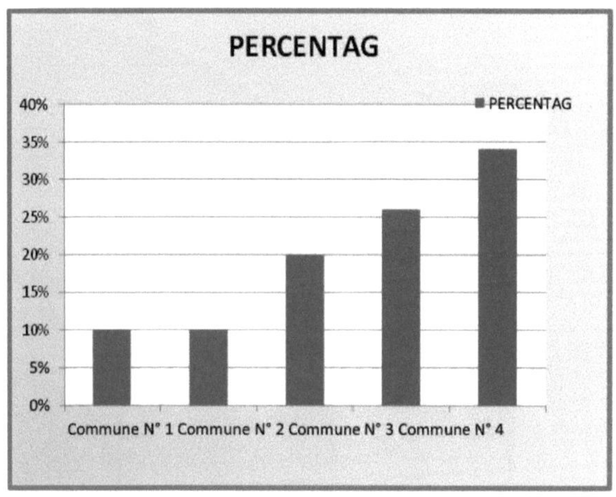

Graph

WOMEN MEN

Group I (12 to 33 years)	86%	90%
Group II (34 to 57 years)	13%	4%
GroupIII (58 and over)	1%	6%

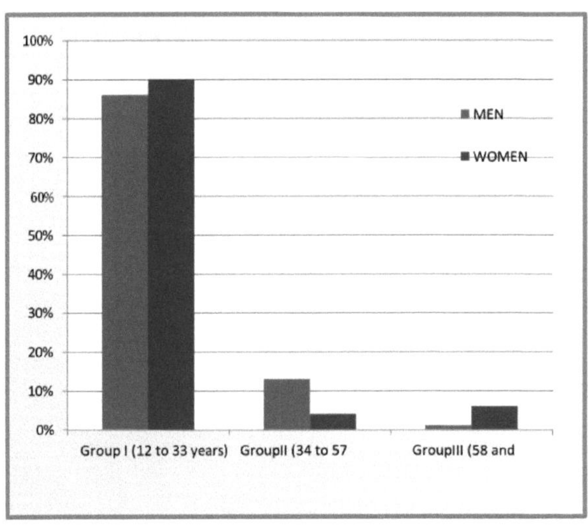

LIST OF NICKNAMES

Nicknames are a special, affectionate and even funny way of naming someone you love and trust. That's why, in a relationship full of love and complicity, it's common for partners to have cute or naughty names for each other. If you want to bring out your most daring and sensual side to surprise someone you are trying to conquer, this article will interest you.

Nicknames for handsome men

We all think that our boyfriend is the most handsome, but also the most special. That's why we want to offer you some original compliments so that you can adapt them to your man. These are the **nicknames for handsome men** that we propose to you:

- Big eyes
- Daddy
- Daddy
- Ricura
- Caramelito
- Cupcake
- Body
- Mr. Sexy
- Playboy
- Dimples
- Curly
- Moreno
- Blond or rubiales
- Negrito
- Pearl
- Skinny or skinny
- Bonbon
- Beautiful
- Nice
- Adonis
- Caraguapa
- Chati
- Pimp

- Mouth
- Guapo
- Fortachón
- Musculitos
- Hottie

Nicknames for nice men

If you want to dedicate a nickname to that special guy so that he understands in sweet words all that he means to you, here are some **cute nicknames for men** that he's sure to love. Which one best defines your partner?

- Bello
- Baby or infant
- Nice
- Kitty
- Life
- My King
- Heart
- Bollito
- My life
- My child
- My sky
- Sweetness
- Monino
- Gift
- Tenderness
- Cheese
- Bolita
- Chicken
- Chispita

Funny nicknames for men

We know that you like to add humour and sparkle to your relationship, that's why we propose some funny nicknames for men that will help you to add spice and grace to your relationship. The most important thing is to be original and to enhance the fun and funny side of your man. Write down these **funny nicknames for men**:

- Pigeon
- Rosebud
- Petit Suisse
- Gherkin
- Flower
- Titi
- Zarigüella
- Chatungo
- Chorbo
- Chief
- Stallion
- Misifú
- Soldier
- Lint
- Mufflers
- Shark
- Crocodile
- Nutty or crazy
- Cuchi cuchi
- Gnome
- Pichín
- Pichurrita
- Churro
- Pocholito
- Pulguita
- My love
- Mochi
- Sausage or sausage
- Currupipi
- Chichipán
- Chichinabo

Tender nicknames for men

Do you like to be tender? If the answer is yes, you probably already know the following list of nicknames. Bring out your sweetest and most tender side to name that special person. Remember: men also like to feel special and loved. Which of these **tender nicknames for men** do you choose?

- Monkey
- Solete
- Dwarf
- Little
- Chiqui or chiquitín
- Churri
- Cuddly toy or cuddly toy
- Cari
- My treasure or my little treasure
- Prince
- Critter
- Sweetheart
- Heaven or heaven
- My love
- Little Lion
- Puppy
- Tadpole
- Teddy Bear
- Birdie

Bold nicknames for men

If you don't want to call that guy in such a classic and traditional way, take a look at the following **cheeky nicknames for men**:

- Playful
- Naughty
- Delight
- Shrew
- Taste
- Mr

- Butt or bum
- Canijo
- Doll
- Nibbles
- Chicken
- Fox
- Little Chief
- Tarzan
- Superman
- Juguetito
- Peleon

In this list, you will also find the best **sex nicknames for men**, so if you are looking for something more racy to spice up your games in bed, choose from this list the most suitable one!

Cute and sweet nicknames for girlfriends

Let's start with some cute nicknames perfect for friends that will show the special bond between you. Here are some of our favourites:

- Baby
- Love
- Sky
- Bichín
- Kookie
- Bug
- Beffi
- Cuca
- Little
- Flaqui
- Gordi
- Caramel
- Muñe (or doll)
- Chiqui
- Heart
- Queen
- Linda

Affectionate nicknames for friends - with meaning!

Do you want some cute nicknames for friends that are funny and at the same time have a specific meaning? If this is your case, don't miss the list below:

- **Hormigui**: perfect nickname for smaller and/or shorter friends.
- **Olive**: also suitable for short or petite friends who are small and beautiful.
- **The earthquake**: the soul of the party, a real nerve that is always coming up with new things.
- **La nervio**: a variant very similar to the previous nickname.
- **Butterfly**: the most beautiful and/or elegant friend or the one who is always full of colour.
- **The loquacious one**: she's the craziest friend of all, but you know that without her nothing would be the same.
- **Princess**: the most delicate friend, the one who makes you laugh with her way of being.
- **Diva**: the diva of them all, that's for sure. It's perfect for that posh friend who makes you laugh with her witticisms and sarcasm.
- **Sol or Solete**: perfect for a radiant and optimistic friend; the one who always brings a smile to your face and knows how to make your day a better one.
- **The party**: she doesn't miss a single one! The friend with this nickname loves going out, having fun, drinking and singing with her group of friends.
- **Teddy**: this nickname is a must, as it is perfect for those tender and sweet friends who are always by your side in bad times.
- **Angelita or angelín**: if you have that typical friend who always behaves well in the eyes of others but when alone is a devil... this nickname is the most appropriate for her.
- **Rici**: perfect for friends with curly hair.
- **The skull**: typical nickname for that tall, thin friend.
- **Kami**: also "kamikaze", this nickname is ideal for that crazy friend who joins any party and gives it all until dawn.

More original and special nicknames for friends

Are you still hungry for more? Don't worry, we offer you some more original and funny nicknames for friends. Which one best defines each of your friends?

- Blonde

- Tessoso
- Kitty
- Puppy
- Sweetness
- Dulcita
- Cuchi
- Petardita
- Amichi
- Crumb
- Cuchi
- Vivi
- Perlite
- Mimi
- Elf
- Strawberry
- Lulu
- Neni

Funny and original nicknames for friends

Want to make your friends laugh out loud? If you're always joking and you want to find the perfect nickname for that special friend... don't miss the following suggestions!

- Adonis
- Cuchufleta
- Chochi
- Gusiluz
- Chicha
- Baby
- Retaco
- Little thing
- Minini
- Bomboncín
- The shots
- Chispita
- Skunk
- Peluchín

- Petit-suisse
- Zombie
- Loba
- Melosa

Non-cheesy nicknames for my girlfriend

It is not easy to find nicknames for your girlfriend that are not cheesy, because nowadays there are very funny, but somewhat embarrassing options that many are reluctant to use. That's why from a COMO we want to start with a few lifelong **nicknames**; enjoy these classic proposals that we have selected for you:

- Love
- Sweetheart
- Sky
- My life
- Heart
- Flower
- Jewel
- Angel
- Preciosa or precioso
- Tender
- Queen
- King
- Treasury
- Bonita or bonito
- Caraguapa
- Bella
- Smurf
- Kookie
- Little

Funny nicknames for your boyfriend or girlfriend

It doesn't always have to be all serious and lovey-dovey; in fact, **funny nicknames for couples** are some of the most popular nowadays. Whether it's for a laugh, to irritate your partner as a joke or to come up with an original nickname, here are some names that are sure to become a private joke that only you and your partner will understand... The fun is guaranteed!

- Chiquitín or chiquitina
- Churri
- Amorcin
- Churrita or churrito
- Baby
- Tarzan
- Tiger or tigress
- Mon cheri
- My cuqui
- Panzita
- Cuchufleta
- Little thing
- Canijilla
- Cheeks
- Pretty face
- Rulitos
- Bichejo
- Rosebud
- Fatty or chubby
- Dwarf
- Chuchurrumín

Original nicknames for boyfriends in Spanish

Are you looking for more nicknames for your boyfriend in Spanish? If so, here are some **very funny and original options for you to** share in the most tender and funny moments.

- Chorba
- Amorcito
- Rabbit
- Amore mio
- Pichurri
- Wren

- Ratita
- My weapon
- Flaqui
- Chatungo
- Cleavage
- Chichi
- Courgette
- Caramelito
- Melon heart
- Palomita
- Salerosa
- Shrew
- Louse
- Hottie
- Chaparrita

Lovely nicknames for a friend

- Sweetness of my heart
- Princess Charming
- Shining star
- Angel of my life
- Radiant flower
- Tender love
- Precious pearl
- Cuddly bear
- Sweet kitty
- Light of my existence
- Charming candy
- Beautiful butterfly
- Lovely Queen
- Unrivalled treasure
- Beautiful smile
- Dream Princess
- Sunshine

- Spark of joy
- Little shining star
- Heart full of love
- Angel of tenderness
- Lovely little flower
- Eternal love
- Glittering pearl
- Loving teddy bear
- Pampered kitten
- Light of hope
- Sweet candy
- Radiant butterfly
- Queen of my heart
- Priceless treasure
- Beautiful friend
- Adorable princess
- Ray of happiness
- Spark of sweetness
- Shining star
- Heart full of tenderness
- Angel protector
- Beautiful flower
- Sincere love
- Unique pearl
- Faithful teddy bear
- Charming kitten
- Light of my life
- Sweet candy
- Adorable butterfly
- Queen of love
- Incomparable treasure
- Beautiful jewel
- Dreamy princess

Cute nicknames for lovers

- Amorcito
- Cielito
- Princess
- Heart
- Teddy Bear
- Baby
- Angelito
- My life
- Little Star
- Kitty
- Butterfly
- My love
- Chiquitín
- My treasure
- Buttercup
- Amado
- Prince
- Sweetness
- Dear
- Lover
- Queen
- Amorcito
- My sky
- Pearl
- My love
- Chiquita
- Friend
- Loving
- Beloved
- Little Prince
- Lucero
- My sun

•Melon heart
•Amorcito
•King
•Amorcito
•Little Princess
•Amorcito
•Butterfly
•Amorcito

Cute nicknames for boyfriend

•Amorcito
•Cielito
•Prince
•Heart
•Teddy Bear
•Baby
•Angelito
•Treasury
•My love
•Churrito
•King
•Butterfly
•Little Star
•Cupcake
•Champion
•Chiquitín
•Sol
•Snail
•Nice
•Sweetness
•Little Prince
•Lover
•Chiquito

- Amiguito
- Sky
- Chiquilín
- Pearl
- Reinito
- Angelito
- Teddy bear
- Kitty
- Bonbon
- Amado
- Chiquito bello
- Melon heart
- My little love
- Tenderness
- Mushroom
- Chiquitito
- Prince Charming
- Angelical
- Shooting star
- Angel face
- Chiquis
- King of my heart
- Amorcito lindo
- Churrito de amor
- Heart of gold
- Charming little guy
- Precious pearl

Cute nicknames for girlfriend

- Angelito
- Princess
- Amorcito
- Cielito

- Little Star
- Heart
- My life
- Buttercup
- Sweetness
- Queen
- Butterfly
- Teddy Bear
- Chiquitita
- Moon
- Pearl
- Principessa
- My sun
- Beloved
- Chispita
- Angel of my life
- Treasury
- Amorcita
- Honey drop
- Flower from my garden
- My love
- Little Princess
- My sky
- Wonder
- Sweet love
- Shooting star
- Melon heart
- My Queen
- Butterfly
- Teddy bear
- Chiquita beautiful
- Lunita
- Precious pearl
- Enchanted Princess
- My radiant sunshine
- Eternal love
- Spark of love
- Angel of love

- Treasury
- Amorcito lindo
- Droplet of love
- Flower of my life
- Love of my heart
- My sweet princess
- Wonderful
- Sweet treasure
- Shining star

GEOPOLITICAL MAP OF THE MUNICIPALITY OF TUMACO

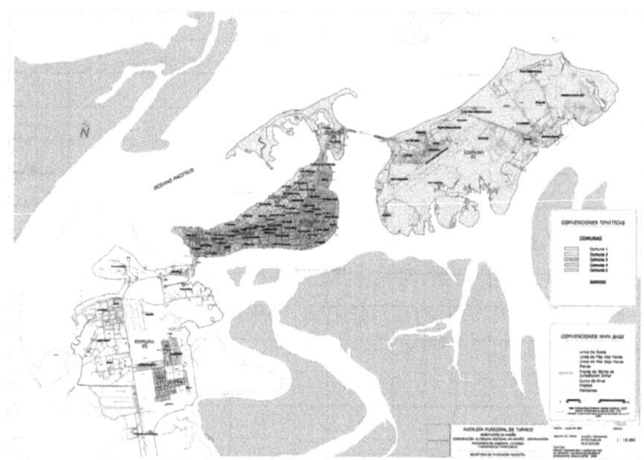

REFERENCES

Abraham, W. (1981). Diccionario de terminología lingüística actual. Madrid: Gredos.
Alvar, Manuel, Manual de dialectología hispánica: el español de España, Barcelona, Ariel, 1996.
Alfonso Martán B., El negro en la poesía negra de Martán Góngora, Popayán, II Seminario sobre Cultura Negra, Universidad del cauca. 1.988.
Alfredo Vanín R., "Palabra Migratoria y Modernidad en el Pacífico". Cartagena de Indias (Colombia). 2000.
Alicia Devalle de R. and Viviana Vega, Una escuela en y para la diversidad. Editorial Aique. Buenos Aires. 1998.
CARLOS ROSELLI P., "El Lenguaje de los Afrocolombianos y su Estudio" América Negra. Javeriana University. Bogotá (Colombia). 1995.
CASTILLO DE LUCAS, A. Spanish nicknames or nicknames, Porto, Imprenta Portuguesa, 1959.
CELA, C. J., El coleccionista de apodos, Madrid, 1947. COSERIU, E., Introducción a la Lingüística. Mexico, UNAM, 1983.
El hombre y su lenguaje: estudios de teoría y metodología lingüística, Madrid, Gredos,1985.
DÍEZ BARRIO, Germán (1995), Motes y apodos, Valladolid, Ed. Castilla.
GARCÍA AGUSTÍN, Ó. (2010), El discurso e institucionalización. Un enfoque sobre el cambio social y lingüístico, Logroño, Universidad de La Rioja.
GONZÁLEZ YANES, María Dolores Emma, Viejos apodos populares. Un estudio sobre las modificaciones introducidas en el lenguaje por la afectividad, La Laguna University de La Laguna: Facultad de Filología, Course: 1993/94 (Unpublished thesis).
HORTENSIA ALAIX DE V., Literatura Popular: Tradición oral en la localidad de El Patía (Cauca), Santafé de Bogotá, COLCULTURA, Tercer Mundo Editores, 1.995.
Agustín Codazzi Institute. Basic Atlas of Colombia. Geo graphic Dissemination Division. 6a. Edition. Bogotá. 1989.
LANGLE, A., Vocabulario, apodos, seudónimos, sobrenombres y hemerografía de la revolución. Mexico, UNAM, 1966.
LE GUERN, M., La metáfora y la metonimia, Madrid, Cátedra, 1976.
LEMUS SANTAMARÍA, V., Apodos, sociedad e individuos: apodos de algunas regiones de Colombia, Tesis de grado, Seminario Andrés Bello, Bogotá, Instituto Caro y Cuervo, 1985.
LIBARDO ARRIAGA C., "La Literatura Oral: Otro Aporte de los Negros",

Cátedra de Estudios Afrocolombianos: GGASA. Bogotá (Colombia) 2002.
MOLINER, M., Diccionario de uso del español, Madrid, Gredos, 1999.
MONTES GIRALDO, J. J. Dialectología general e hispanoamericana: orientación teórica, metodológica y bibliográfica, third edition, Santafé de Bogotá, CARO Y CUERVO, 1995. -Motivación y creación léxica en el español de Colombia, Bogotá, Instituto Caro y Cuervo, 1983.
RAMÍREZ MARTÍNEZ, J. (2003), Los sobrenombres y su aprovechamiento educativo: Sobre los apodos en el Valle Medio del Iregua, Madrid, UNED (Unpublished doctoral thesis, in process of publication).
RAMÍREZ MARTÍNEZ, J. (2005a), "Aprovechamiento educativo y didáctico de los apodos del Campo de Cartagena", Revista Murciana de Antropología: I Congreso Etnográfico del Campo de Cartagena, 2003, Murcia, University of Murcia.
RAMÍREZ MARTÍNEZ, J. and RAMÍREZ GARCÍA, R. (2005b), "Los apodos: Identidad,memoria y creatividad literaria", El Descubrimiento Pendiente de América Latina Diversidad de saberes en diálogo hacia un proyecto integrador (Actas-Memorias del Ier. Foro Latinoamericano: Memoria e Identidad), Montevideo, Signo Latinoamericano.- UNESCO.
REAL ACADEMIA ESPAÑOLA (1992), Diccionario de la lengua española. Madrid, Real Academia Española.
SÁNCHEZ, M., "El apodo como fenómeno sociolingüístico en el Perú", in lenguaje y Ciencias, Universidad de Trujillo, vol. 20, Number 2, Trujillo (Peru), 1980.
SARTOR, M. "El apodo como creación semántica: ensayo de antroponimia", in Anales del Instituto de Lingüística, T. CIV, Mendoza (Argentina), 1988.
ULLMANN, S., Semántica: Introducción a la ciencia del significado, Madrid, Aguilar, 1965.
Manual de dialectología hispánica: el español de España, Barcelona, Ariel, 1996.
Alfonso Martán B., El negro en la poesía negra de Martán Góngora, Popayán, II Seminario sobre Cultura Negra, Universidad del cauca. 1.988.
Alfredo Vanín R., "Palabra Migratoria y Modernidad en el Pacífico". Cartagena de Indias (Colombia). 2000.
Alicia Devalle de R. and Viviana Vega, Una escuela en y para la diversidad. Editorial Aique. Buenos Aires. 1998.
CARLOS ROSELLI P., "El Lenguaje de los Afrocolombianos y su Estudio" América Negra. Javeriana University. Bogotá (Colombia). 1995.

CASTILLO DE LUCAS, A. Spanish nicknames or nicknames, Porto, Imprenta Portuguesa, 1959.

CELA, C. J., El coleccionista de apodos, Madrid, 1947. COSERIU, E., Introducción a la Lingüística. Mexico, UNAM, 1983.
El hombre y su lenguaje: estudios de teoría y metodología lingüística, Madrid, Gredos,1985.

HORTENSIA ALAIX DE V., Literatura Popular: Tradición oral en la localidad de El Patía (Cauca), Santafé de Bogotá, COLCULTURA, Tercer Mundo Editores, 1.995.
Agustín Codazzi Institute. Basic Atlas of Colombia. Geo graphic Dissemination Division. 6a. Edition. Bogotá. 1989.
LANGLE, A., Vocabulario, apodos, seudónimos, sobrenombres y hemerografía de la revolución. Mexico, UNAM, 1966.
LE GUERN, M., La metáfora y la metonimia, Madrid, Cátedra, 1976.
LEMUS SANTAMARÍA, V., Apodos, sociedad e individuos: apodos de algunas regiones de Colombia, Tesis de grado, Seminario Andrés Bello, Bogotá, Instituto Caro y Cuervo, 1985.
LIBARDO ARRIAGA C., "La Literatura Oral: Otro Aporte de los Negros", Cátedra de Estudios Afrocolombianos: GGASA. Bogotá (Colombia). 2002.
MONTES GIRALDO, J. J. Dialectología general e hispanoamericana: orientación teórica, metodológica y bibliográfica, third edition, Santafé de Bogotá, CARO Y CUERVO, 1995.
-Motivación y creación léxica en el español de Colombia, Bogotá, Instituto Caro y Cuervo, 1983.
SÁNCHEZ, M., "El apodo como fenómeno sociolingüístico en el Perú", in lenguaje y Ciencias, Universidad de Trujillo, vol. 20, Number 2, Trujillo (Peru), 1980.
SARTOR, M. "El apodo como creación semántica: ensayo de antroponimia", in Anales del Instituto de Lingüística, T. CIV, Mendoza (Argentina), 1988.
ULLMANN, S., Semántica: Introducción a la ciencia del significado, Madrid, Aguilar, 1965.
https://reportedelectura.org/apodos https://www.bing.com/search?

TABLE OF CONTENTS

Printed by Books on Demand GmbH, Norderstedt / Germany